Cell Transformation

NATO ASI Series
Advanced Science Institutes Series

A series presenting the results of activities sponsored by the NATO Science Committee, which aims at the dissemination of advanced scientific and technological knowledge, with a view to strengthening links between scientific communities.

The series is published by an international board of publishers in conjunction with the NATO Scientific Affairs Division

A	**Life Sciences**	Plenum Publishing Corporation
B	**Physics**	New York and London
C	**Mathematical and Physical Sciences**	D. Reidel Publishing Company Dordrecht, Boston, and Lancaster
D	**Behavioral and Social Sciences**	Martinus Nijhoff Publishers
E	**Engineering and Materials Sciences**	The Hague, Boston, and Lancaster
F	**Computer and Systems Sciences**	Springer-Verlag
G	**Ecological Sciences**	Berlin, Heidelberg, New York, and Tokyo

Recent Volumes in this Series

Volume 89—Sensory Perception and Transduction in Aneural Organisms
edited by Giuliano Colombetti, Francesco Lenci, and Pill-Soon Song

Volume 90—Liver, Nutrition, and Bile Acids
edited by G. Galli and E. Bosisio

Volume 91—Recent Advances in Biological Membrane Studies: Structure
and Biogenesis, Oxidation and Energetics
edited by Lester Packer

Volume 92—Evolutionary Relationships among Rodents: A
Multidisciplinary Analysis
edited by W. Patrick Luckett and Jean-Louis Hartenberger

Volume 93—Biology of Invertebrate and Lower Vertebrate Collagens
edited by A. Bairati and R. Garrone

Volume 94—Cell Transformation
edited by J. Celis and A. Graessmann

Series A: Life Sciences

Cell Transformation

Edited by

J. Celis
Aarhus University
Aarhus, Denmark

and

A. Graessmann
Institute for Molecular Biology
Free University of Berlin
Berlin, Federal Republic of Germany

Plenum Press
New York and London
Published in cooperation with NATO Scientific Affairs Division

Proceedings of a NATO Advanced Study Institute/FEBS/
Gulbenkian Foundation Summer School on
Cell Transformation,
held September 2–12, 1984,
in Sintra-Estoril, Portugal

Library of Congress Cataloging in Publication Data

NATO Advanced Study Institute/FEBS/Gulbenkian Foundation Summer
   School on Cell Transformation (1984: Sintra and Estoril, Portugal)
   Cell transformation.

   (NATO ASI series. Series A, Life sciences; v. 94)
   "Proceedings of a NATO Advanced Study Institute/FEBS/Gulbenkian
Foundation Summer School on Cell Transformation, held September 2–12,
1984, in Sintra-Estoril Portugal"—T.p. verso.
   "Published in cooperation with NATO Scientific Affairs Division."
   Includes bibliographies and index.
   1. Carcinogenesis—Congresses. 2. Oncogenes—Congresses. 3.
Cancer—Genetic aspects—Congresses. I. Celis, J. E. (Julio E.) II.
Graessmann, A. III. NATO Advanced Study Institute. IV. Federation of Euro-
pean Biochemical Societies. V. Fundáçao Calouste Gulbenkian. VI. North
Atlantic Treaty Organization. Scientific Affairs Division. VII. Title. VIII. Series.
[DNLM: 1. Cell Transformation, Neoplastic—congresses. 2. Cell Transforma-
tion, Viral—congresses. 3. Gene Expression Regulation—congresses. QZ 202
N2785c 1984]
RC268.5.N345   1984   616.099′4071   85-16923
ISBN-13: 978-1-4684-5011-8   e-ISBN-13: 978-1-4684-5009-5
DOI: 10.1007/978-1-4684-5009-5

© 1985 Plenum Press, New York
Softcover reprint of the hardcover 1st edition   1985
A Division of Plenum Publishing Corporation
233 Spring Street, New York, N.Y. 10013

PREFACE

This volume is based on the proceedings of a NATO/FEBS/Gulbenkian
Foundation sponsored Summer School held in September 1984 in Sintra-
Estoril, Portugal.

Given the accelerated growth of knowledge in the field of cell
transformation, it seemed timely to hold a summer school to discuss
current developments in this area of biology as well as to evaluate
emerging technology. The *first* article in this volume gives an
evaluation of the various cellular systems to study neoplasia.
Their properties as well as advantages and disadvantages are discussed. The *second* section deals with the role of oncogenes in cell
transformation. Particular emphasis is given to the question of
whether activated proto-*onc* genes are cancer genes and to the functions of oncogene products. The *third* part is dedicated to viruses
and includes articles on papova viruses, Epstein-Barr virus, adenovirus, parvo viruses and HTLV. The *fourth* part deals with gene expression in normal and transformed cells while the *concluding* section considers various aspects of gene regulation in eukaryotic
cells.

We wish to express our appreciation to Dr. Maria C. Lechner who provided valuable advice and help concerning the organization of this meeting. We are also indebted to Ms. Lisbeth Heilesen and Ms. Anne Mette Lygaard for typing the manuscripts and for their outstanding administration of the meeting.

J.E. Celis

A. Graessmann

February 1985

CONTENTS

NEOPLASTIC TRANSFORMATION SYSTEMS

1. Neoplastic Transformation Systems - Their Use In Studying Carcinogenesis.....................................   1
      A. Sivak & A.S. Tu

ONCOGENES

2. Are Activated Proto-*onc* Genes Cancer Genes?.............   21
      P.H. Duesberg, M. Nunn, N. Kan, D. Watson,
      P.H. Seeburg & T. Papas

3. Immunoglobulin Genes, Oncogenes, and Human B-Cell
   Tumors...   65
      P.C. Nowell & C.M. Croce

4. The Functions of Oncogene Products.....................   79
      T. Hunter

5. Identification and Localization of Phosphoproteins in
   v-*onc* Transformed Fibroblasts by Means of Phosphotyrosine Antibodies...................................   97
      P.M. Comoglio, D. Cirillo, M.F. Di Renzo,
      R. Ferracini, F.G. Giancotti, S. Giordano,
      L. Naldini, G. Tarone & P.C. Marchisio

VIRAL TRANSFORMATION

6. The Transformation Capacity of Early SV40 DNA Fragments...   113
      A. Graessmann & M. Graessmann

7. The Transforming Genes of Polyoma Virus................ 127
   M. Rassoulzadegan & F. Cuzin

8. Papova Viruses and Cancer Genes........................ 135
   C. Streuli & B.E. Griffin

9. Epstein-Barr Virus and Immortalisation of Epithelial
   Cells.. 157
   B.E. Griffin

10. Functional Domains of Purified Adenovirus Type C
    E1A Proteins.. 167
    B. Krippl, B. Ferguson, N. Jones, M. Rosenberg,
    & H. Westphal

11. Parvoviruses and Cancer............................... 175
    B. Hirt

12. HTLV in Adult T Cell Leukemia and Acquired Immune
    Deficiency Syndrome................................... 185
    P.S. Sarin

GENE EXPRESSION IN NORMAL AND TRANSFORMED CELLS

13. Construction of Protein Databases for Comparison of
    Normal and Transformed Cells.......................... 209
    J.I. Garrels & B.R. Franza, Jr.

14. Cyclin (PCNA) is a Component of the Pathway(s) Leading
    to DNA Replication and Cell Division: A Role in DNA
    Replication?.. 223
    J.E. Celis & A. Celis

REGULATION OF GENE EXPRESSION

15. Regulation of Gene Expression in Developmental and
    Oncogenic Processes: The Albumin Alpha-Fetoprotein
    Locus in Mammals...................................... 239
    J.M. Sala-Trepat, A. Poliard, I. Tratner, M. Poiret,
    M. Gomez-Garcia, A. Gal, J.L. Nahon, & M. Frain

16. Transcription Control in Eucaryotes-Enhancers and
    Promoters.. 267
        B. Bourachot, P. Herbomel & M. Yaniv

17. Controls of Gene Expression in Chemical Carcinogenesis:
    Role of Cytochrome P450 Mediated Mono-Oxygenases........ 285
        M.C. Lechner

Contributors... 313

Index.. 317

NEOPLASTIC TRANSFORMATION SYSTEMS - THEIR USE IN STUDYING CARCINOGENESIS

Andrew Sivak & Alice S. Tu

Biomedical Research and Technology Section
Arthur D. Little, Inc., Acorn Park
Cambridge, Massachusetts 02140, USA

INTRODUCTION

The cellular systems to study neoplasia essentially stem from two sources. One is the observation that one can induce alterations in cellular phenotype in a culture of cells infected with tumorigenic DNA viruses (1). The second is the finding of Berwald and Sachs in 1963 (2,3) that early passage Syrian hamster embryo cells exhibited clonal morphology not seen in untreated cultures following exposure to a chemical carcinogen. While the work of Earle and his associates (4) beginning in the nineteen thirties had demonstrated changes in cell cultures treated with carcinogens, it was the protocol and results reported in 1963 (2,3) that provided a means to obtain quantitative and reproducible findings of morphological transformation of mammalian cells induced by chemical carcinogens. Over the past two decades a considerable variety of systems have been described to study neoplastic transformation. Of these, several have been shown to have value as bioassays for the identification of carcinogens. The properties as well as the advantages and disadvantages of these assays have been reviewed in depth recently along with a presentation of the available data base on tested chemicals (5-9). Table 1 lists these assay types along with some basic characteristics.

1

Table 1. *Neoplastic Transformation Assays*

Cell System	Cell Type	Assay Interval	Assay Type	Quantitation* Toxicity	Transformation
Syrian hamster embryo	Strain	8-10 days	Clonal	+	+
Mouse BALB/c-3T3	Line	30 days	Focus	±	+.
Mouse C3H-10T½	Line	45 days	Focus	±	+
Syrian hamster embryo + simian adenovirus	Strain	25-30 days	Focus	±	+
Fischer rat embryo + Rauscher leukemia virus	Line	6-8 weeks	Focus	±	-

* The evaluation of quantitation is + = unequivocal, ± = ambiguous, - = not possible in terms of being able to obtain measured frequencies of toxicity or transformation.

CONTEMPORARY TRANSFORMATION ASSAYS

The Syrian hamster embryo (SHE) transformation assay has several clearly desirable characteristics. The cells are diploid and in early *in vitro* passage. The assay interval is relatively short and numerical results can be obtained in a quantitative manner because it is a clonal assay. However, as several studies with this assay have revealed, it is extremely sensitive to environmental variables such as fetal calf serum source and specific embryo cell pool (10,11). Moreover, the wide variation in morphology of transformed clones makes unambiguous scoring a problem.

Two assays use cell lines established from mouse embryo cultures (BALB/c-3T3, C3H-10T½) by rigorous subconfluent passaging of cultures through a crisis stage until they reached stable characteristics of saturation density and cloning efficiency (12-15). Both lines are heteroploid with modes near the tetraploid range, exhibit

a strong density dependent inhibition of cell division and can be
induced by chemical and physical carcinogens as well as oncogenic
viruses to exhibit morphologically altered foci against an untransformed monolayer (16). The lines differ from each other in terms of
variability in response to serum source and sensitivity to carcinogens in the bioassay, with the BALB/c-3T3 line showing less variable
response to different serum lots as well as being more sensitive

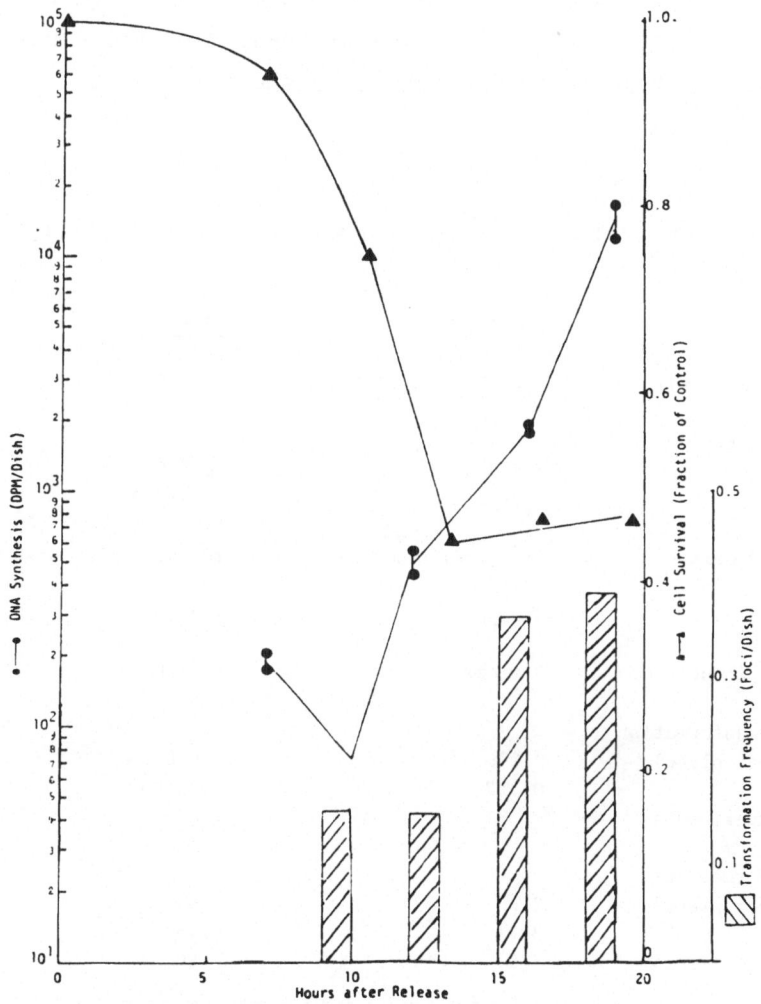

Cell Cycle Dependency of Cytotoxicity and Transformation of C3H-10T 1/2
Cells Treated with MNNG (1 μg/ml - 10⁵ Plating Density)

Figure 1.

in transformation response (17). The C3H-10T½ line has a substantially lower spontaneous transformation frequency than BALB/c-3T3 cells. Another striking difference is the cell cycle specificity of the C3H-10T½ cells in transformation by potent direct acting carcinogens such as N-methyl-N'-nitro-N-nitroso-guanidine (Figure 1). The BALB/c-3T3 cells do not exhibit this cell cycle specificity. In contrast to the SHE assay, the scoring of transformation is less ambiguous, with transformed foci falling into generally recognizable classes. A list of key properties of these two cell lines and their behavior in transformation assays is shown in Table 2.

Table 2. *Growth Properties and Transformation.* Characteristics of BALB/c-3T3 and C3H-10T½ cells

	BALB/c-3T3	C3H-10T½
Cloning Efficiency (%)	27-51	15-29
Saturation Density		
$x10^6$/60mm dish	1.5-2.4	0.5-0.9
$x10^4$/cm^2	7.2-11.4	2.4-4.3
Assay Plating Density	10^4 cells/60mm dish	10^4 cells/60mm dish
Treatment Time	3 days	1 day
Post Treatment Incubation	4 weeks	6 weeks
Spontaneous Transformation		
Mean (foci per plate)	0.276	0.0007
Variance	0.312	0.00003
Number of experiments	87	59
Induced Transformation*		
Mean (foci per plate)	2.6	0.61
Variance	3.2	0.16
Number of experiments	75	60

* 2 μg/ml and 5 μg/ml 3-methylcholanthrene treatment for BALB/c-3T3 and C3H-10T½ cells, respectively.

Another neoplastic transformation assay using cell populations derived from the cell line BHK-21 has been described (18). This procedure measures the increase in the ability of carcinogen treated cells to grow in an anchorage independent condition (soft agar) compared to untreated cells. While a modest data base was developed rapidly (18-20), issues have arisen with respect to the selection and stability of the target cell populations as well as a consideration of what the assay was actually measuring. Evidence has been offered that the occurrence of anchorage independence in these cells may be the result of a single step mutation like process (21). In any event, there does not appear to be substantial activity with this assay procedure at the present time.

Of the assays being evaluated as tools for carcinogen identification, two employ viruses. The Syrian hamster embryo - simian adenovirus (SA-7) assay measures an enhancement of a virus-induced focal transformation response that occurs after carcinogen exposure (22). Although this assay is one that seems to measure the carcinogen induced modulation of transformation caused by virus rather than directly by the carcinogen, the uniformity of transformed foci and relative ease of scoring makes this assay a facile one to perform. The specific role of the carcinogen has not been definitively shown, however, there is convincing information that DNA damage is closely associated with the enhancement of virus-induced transformation, suggesting that carcinogen treatment may increase the number of viral integration sites (23).

The Fischer rat embryo - Rauscher leukemia virus assay was first described by Freeman *et al.* in 1970 (24). This assay employs rat cells that carry an apparently unproductive infection of Rauscher leukemia virus that renders the cells sensitive to transformation by chemical carcinogens. A substantial data base has been developed using the focus assay (7). However, since this assay requires sub-

culturing after treatment, it is not possible to obtain a determination of frequency of transformed foci that can be related quantitatively to the original target cell population. A modification of
the assay to determine anchorage independent survival of carcinogen-
treated cells (25) is under evaluation.

In addition to these transformation systems that largely employ
fibroblast as target cells, a number of culture systems have been
described that utilize epithelial cells. Transformation responses
have been demonstrated with mouse epidermal keratinocytes (26,28).
rat tracheal epithelial cells (29), rat hepatocytes (30) and mouse
mammary cells (31) with sufficient frequency to define these systems
as effective tools to examine organ specific transformation events.
However, because these systems are so time and energy intensive,
their applicability in a general carcinogen identification program
is limited.

FACTORS INFLUENCING ASSAY

Metabolism - For each of these assays, a substantial number of
chemicals have been tested and the correlation to *in vivo* carcinogenicity for aromatic hydrocarbons, direct-acting agents and metals
is good (7,9,17). One of the major unsolved drawbacks of all of the
assays now being considered for carcinogen identification is the
narrow metabolic range of the target cells.

The absence of response with many aromatic amines, nitrosamines and
other procarcinogens requiring metabolic conversion has been addressed with the application of exogenous metabolic activation systems.
Essentially, two vehicles have been employed to add metabolic capability to the systems: isolated hepatocytes from various rodent
species (mouse, rat, hamster) and 9,000xg supernatants (S-9 fractions) from liver tissue of these same animals.

Table 3. *Transformation Studies with C3H-10T½ Cells and Rat Hepatocytes#*

	Control	Cyclophosphamide*		AA*	MCA*
		5.0	10.0	20.0	5.0
With Hepatocytes					
Type III (foci/plate)	0.05 ± 0.05	0	0.53 ± 0.22	0	**
Type II + III (foci/plate)	0.10 ± 0.07	0.05 ± 0.05	0.53 ± 0.22	0.15 ± 0.11	**
Cloning Cytotoxicity (T/C)		0.66	0.69	1.00	0.03
No Hepatocytes					
Type III (foci/plate)	0	0.05 ± 0.05	0.05 ± 0.05	0	0.10 ± 0.07
Type II + III (foci/plate)	0	0.40 ± 0.11	0.35 ± 0.13	0	0.40 ± 0.15
Cloning Cytotoxicity (T/C)		1.20	0.95	1.10	0.03

* Doses in µg/ml; AA = 2-Aminoanthracene, MCA = 3-Methylcholanthrene

** Dense with, no defined foci

Target cells were plated after a seven-day recovery period

The available data suggest that there may not be a generalized
solution to the problem. Depending on the assay system and the
specific chemical, it has been observed that either hepatocytes or
S-9 fractions can be more effective in providing the necessary
metabolic activation to induce transformation (32,33). Moreover,
the induction of transformation does not appear to correlate with
the production of cytotoxicity by the metabolic activation system
(Table 3).

Tumor promotion - An advantage often cited for the use of cellular
neoplastic transformation systems is that the response measured
bears resemblance to the biological sequence of events occurring
in carcinogenesis *in vivo*. While the analogy is not ideal, at
least four of the transformation systems provide a means to measure
a tumor promotion response like those found *in vivo*. Although
studies as early as 1967 (34) indicated that a tumor promoting
response could be elicited in cell culture systems, the finding
from Heidelberger's laboratory (35) in the C3H-10T½ system was the
first clear demonstration of a two-stage experiment that was model-
ed after the classical mouse skin initiation-promotion protocols.
Later studies by a number of other laboratories explored the details of the response and some of the limiting factors in this
cell system (36,37). Demonstrations of transformation enhancing
activity were also reported for two other cell line transformation
systems (BALB/c-3T3 (36) and Fischer rat embryo cell-Rauscher
virus (39).

Recently, several investigators have shown that the SHE system
responds to tumor promoting agents (40-42). This later finding is
of considerable interest since the SHE assay is a clonal procedure
and transformation response is one of individual clonal behavior.
This is in contrast to the focus assays where the transformation
response in tumor promotion studies is a complex interaction be-

Table 4. *Transformation of SHE Cells by B(a)P in Various Fetal Calf Serum Lots*

Supplier	Lot#	%C.E.	Control	1 µg/ml	3 µg/ml	5 µg/ml
Hyclone	100347	24.6	0/1036	0/951	1/114	-
Hyclone	100348	16.7	0/747	1/806	1/668	-
Gibco	31K4310	16.7	0/748	1/513	1/568	1/568
Gibco	31N7013	14.3	0/638	1/359	0/398	1/410
MAB	2A037	25.0	0/1129	0/688	1/815	0/803
MAB	2A051	25.7	0/1148	1/966	2/1024	1/1018
Reheis	V55503	28.7	0/1283	1/683	0/866	-
Reheis	U54012	30.0	0/1348	4/844	8/903	11/907

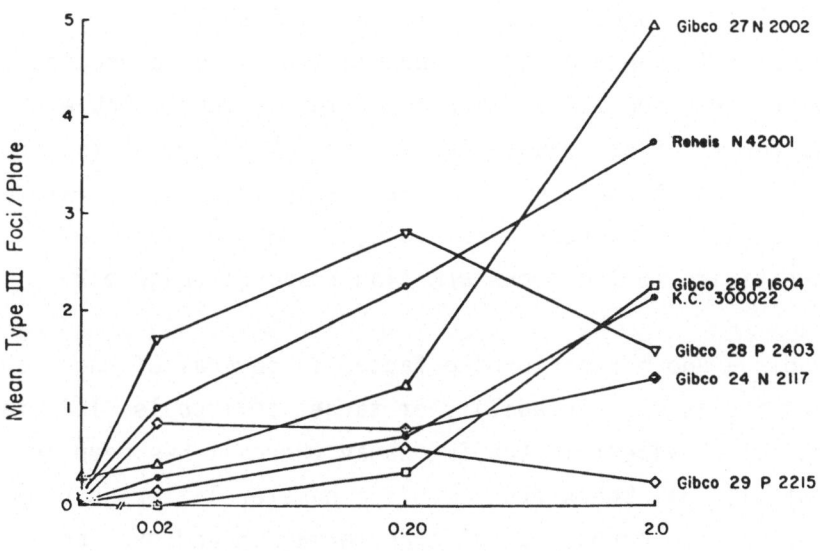

Transformation of BALB/c **3T3 Cells** by MCA in Various Fetal Calf Serum Lots

Figure 2.

tween transformed cells and the monolayer of untransformed cells. Thus, the enhancement of the transformation in several different systems by tumor promoters offers the possibilities of studying the action of promoters at the cellular level as well as providing a potential means to increase the sensitivity of the assays.

Serum Supplement - While the use of exogenous metabolic activation systems and tumor promoters may help to increase the specificity and sensitivity of the transformation assays, there remain two factors that represent continuing problems and are inherent in the assays. One of these factors is the sensitivity of the transformation assay to the specific fetal calf serum lot. The SHE and C3H-10T½ assays seem to be exceptionally sensitive to the lot of serum employed. The reasons for this are not clear and no components of fetal calf serum have been specifically identified that could account for the variability that is seen. The practical outcome of this is that sera must be screened empirically to find a suitable lot to support transformation. The ability of a serum to support clonal growth of a cell population is not a predictive parameter for supporting a transformation response. Table 4 shows results for a set of sera tested in the SHE system. Although the BALB/c-3T3 transformation system is less exacting with respect to its response to different serum lots, it does exhibit some serum lot sensitivity to the transformation response (Figure 2).

Cell Pools - Another perturbing factor in several of the transformation assays is the variability of target cell pools. This factor is especially evident in the SHE assay where it has been found that not all cell pools from specific hamster litters can be induced to undergo morphological transformation and even cell pools from individual pups in a single litter can vary in their transformation response (11). Although these results have been repeated in several laboratories, the basis for this finding with respect

to identification of specific cellular properties that allow or
prevent transformation remains unknown.

A similar observation has been made with BABL/c-3T3 cells with
respect to the heterogeneity of response (43) of target cell clones.
The mechanistic reason for this variation is largely unknown. However, a number of factors including chromosome constitution, in-
ducibility of gene mutations, transformation with Kirsten sarcoma
virus or differences in metabolic profiles for polycyclin aromatic
hydrocarbon carcinogens have been ruled out as being contributory
to the ability of a cell to be transformed by chemical or physical
carcinogen. A general observation that has been made with populations of this particular cell line is that the frequency of carcinogens induced transformation appears to be coupled to the spontaneous transformation frequency (Table 5).

Table 5. *Transformation of BALB/c-3T3 Clones of Differing Sponta-*
neous Transformation Frequencies

Clone	Type III Foci/Total Number of Plates	
	Control	MCA (2µg/ml)
A31	0/20	0/20
A31-1-11	0/21	7/20
A31-1-13	4/18	58/19
A31-1-13-5	0/16	1/12
A31-1-13 (cumulative)*	5/20	52/20

*Based on 20 plates per set. Data from a three year observaton period.

WHAT DO NEOPLASTIC TRANSFORMATION ASSAYS MEASURE?

The rationale usually given for the use of cellular neoplastic
transformation assays to identify carcinogens and/or to study the
process of carcinogenesis is that the phenomenon of induced altered morphology in culture is coupled with the acquisition of anchorage independent growth *in vitro* and oncogenicity *in vivo* and this

correlation provides a basis for comparison to events in *in vivo* carcinogenesis. Beyond the evident absence in the transformation systems of immunological or hormonal factors, which have potent influences on the outcome of the neoplastic process *in vivo*, the problems with the correlation between morphological transformation and other markers for the tumor phenotype (plasminogen activator, anchorage independent growth, tumor formation *in vivo*) suggests that this rationale should be reexamined. A series of studies by Kennedy and Little (42-44) with the C3H-10T½ system have raised questions about the genetic nature of the inducing event in transformation, especially with established cell lines.

A critical factor that merits consideration is the physiological sensitivity of the transformed phenotype. The extensive studies of Rubin (47-50) have documented that population density as well as inorganic ion level of the medium, especially hypertonicity, have dramatic effects on the expressed phenotype. These findings along with the large heterogenicity among parallel populations of similarly derived transformed populations in terms of anchorage independence, oncogenicity *in vivo* and rates of progression of regression among these characteristics suggest that the morphologically altered phenotype observed in transformed cell cultures may be a very early and unstable step in the neoplastic process. The work of Newbold (49) with Syrian hamster embryo cells confirms this heterogenicity of response of these mixed cell populations to carcinogen treatment and offers support for the interesting view that induction of immortality of a cell population is the initial rate-limiting step in neoplastic transformation that may induce a progressive response in a fraction of the treated populations. In subsequent work (50), it was demonstrated that transfection with H-ras DNA to produce morphologically transformed foci was only possible in immortalized hamster populations suggesting further a potential key role for this property in neoplastic transformation.

However, the variability of transformation responses induced by carcinogens in immortal cell lines indicates that there are additional factors that need to be considered in explaining the mechanisms by which cells are converted from exhibiting an ordered and controlled growth pattern to one of cellular disorientation and invasion into a surrounding controlled growth pattern.

The involvement of expressed oncogenes in carcinogen transformed cells has been examined in a limited number of different populations, and for transformed clones of C3H-10T½ cells, the transfection of NIH-3T3 cells by DNA from transformed foci is not a general finding (Table 6).

Table 6. *Oncogene Transfection Studies*

Cell Line	Transforming Chemical	Transfection Efficiency*	Restriction Enzyme Sensitivity		
			B am H1	XHoI	SA1I
C3H-10T½	–	1			
C3H-10T½ (MCA5)	3-MC**	5,6,8	7	4	8
C3H-10T½ (MCA 16-5)	3-MC	73,58	14,25	33	50
C3H-10T½ (MCA 66,ACL 6)	3-MC	0			
C3H-10T½ (PBI)	BP**	0			
C3H-10T½ (F-2407-NQO)	NQO**	0			

* Foci per 75μg DNA per 1.5×10^{6} NIH-3T3 target cells
Ref. Shih et al., PNAS, 76:5714 (1979); Shilo and Weinberg, Nature, 289,607 (1981).

** 3-MC-3-Methylcholanthrene; BP-Benzo(a)pyrene; NQO-Nitroquinoline-N-oxide

For purposes of identifying environmental chemicals that may have potentially detrimental biological activity, these observations that raise unresolved questions about mechanism may be relatively inconsequential. However, the use of morphological transformation in cell culture systems as a means to study mechanisms of carcinogenesis as applied to *in vivo* events requires careful consider-

ation and appropriate validation of the systems with special attention to the role of environmental variable that can alter the
phenotypic expression of transformation.

REFERENCES

1) TODARO, G.J. & GREEN, H. (1966). Cell growth and the initiation
of transformation by SV40. Proc. Nat. Acad. Sci. USA, 55, 302-
308.

2) BERWALD, Y. & SACHS, L. (1963). *In vitro* transformation with
chemical carcinogens. Nature (London), 200, 1182-1184.

3) BERWALD, Y. & SACHS, L. (1965). *In vitro* transformation of
normal cells to tumor cells by carcinogenic hydrocarbons. J.
Nat. Cancer Inst., 35, 641-661.

4) EARLE, W.R. & VOEGTLIN, C. (1938). The mode of action of
methylcholanthrene on cultures of normal tissues. Am. J. Cancer, 34, 373-390.

5) SIVAK, A. (1969). *In vitro* transformation - Overview and status.
J. Assoc. Off. Anal. Chem., 62, 889-899.

6) SIVAK, A. & TU, A. (1982). Transformation of somatic cells in
culture. In: "Mutagenicity: New Horizons in Genetic Toxicology",
J. Heddle, ed., Academic Press, Inc., New York, pp. 143-169.

7) HEIDELBERGER, C., FREEMAN, A.E., PIENTA, R.J., SIVAK, A.,
BERTRAM, J.S., CASTO, B.C., DUNKE, V.C., FRANCIS, M.W.,
KAKUNAGA, T., LITTLE, J.B. & SCHECHTMAN, L.M. (1983). Cell
transformation by chemical agents - a review and analysis of
the literature. Mutation Res., 114, 283-385.

8) MEYER, A.L. (1983). *In vitro* transformation assays for chemical carcinogens. Mutation Res., 115, 323-338.

9) SIVAK, A. & TU, A. (1984). In: "Handbook of Carcinogen Testing".
H. Milman, ed., E. Weisburger Noyes Publications, Mill Road,
New Jersey.

10) PIENTA, R.J., POILEY, J.A. & LEBHERZ III, W.B. (1977). Morphological transformation of early passage golden Syrian hamster

embryo cells derived from cryopreserved primary cultures as a
reliable *in vitro* bioassay for identifying diverse carcinogens.
Int. J. Cancer, 19, 642-655.

11) DIPAOLO, J.A. (1980). Quantitative *in vitro* transformation of
Syrian golden hamster embryo cells with the use of frozen
stored cells. J. Natl. Cancer Inst., 64, 1485-1489.

12) TODARO, G.J. & GREEN, H. (1963). Quantitative studies of the
growth of mouse embryo cells in culture and their development
into established lines. J. Cell Biol., 17, 299-313.

13) AARONSON, S.A. & TODARO, G.J. (1968). Development of 3T3-like
lines from BABL/c mouse embryo cultures: transformation susceptibility to SV-40. J. Cell. Physiol., 72, 141-148.

14) REZNIKOFF, C.A., BRANKOW, D.W. & HEIDELBERGER, C. (1973).
Establishement and characterization of a cloned line of C3H
mouse embryo cells sensitive to post confluence inhibition of
division. Cancer Res., 33, 3231-3238.

15) REZNIKOFF, C.A., BRANKOW, D.W. & HEIDELBERGER, C. (1973).
Quantitative and qualitative studies of chemical transformation of cloned C3H mouse embryo cells sensitive to postcon-
fluence inhibition of cell division. Cancer Res., 33, 3239-
3249.

16) KAKUNAGA, T. (1973). A quantitative system for assay of malignant transformation by chemical carcinogens using a clone
derived from BABL/3T3. Int. J. Cancer, 12, 463-473.

17) TU, A. & SIVAK, A. (1982). Comparison of oncogenic potential
of carcinogen-induced transformed populations of BALB/c-3T3
and C3H-10T½ cells. Proc. Am. Assoc. Cancer Res., 23, 75.

18) DIMAYORCA, G., GREENBLATT, M., TRAUTHEN, T., SOLLER, A. &
GIORDANO, R. (1973). Malignant transformation of BHK 21 clone
13 cells *in vitro* by nitrosamines; A conditional state. Proc.
Natl. Acad. Sci., 70, 46-49.

19) ISHII, Y., ELLIOT, J.A., MISHRA, N.K. & LIEBERMAN, M.W. (1977).
Quantitative studies of transformation by chemical carcinogens and ultraviolet radiation using a subclone of BHK 21

clone 13 Syrian hamster cells. Cancer Res., 37, 2023-2029.

20) PURCHASE, I.F.H., LONGSTAFF, W., STYLES, J.A., ANDERSON, D., LEFEVRE, P.A. & WESTWOOD, F.R. (1978). An evaluation of six short-term tests for detecting organic chemical carcinogens. Br. J. Cancer, 37, 873-959.

21) BOUCK, N. & DIMAYORCA, G. (1982). Chemical carcinogens transform BHK cells by inducing a recessive mutation. Mol. Cell Biol., 2, 97-105.

22) CASTO, B.C., PIECZYSNKL, W.J. & DIPAOLO, J.A. (1974). Enhancement of adenovirus transformation of hamster cells *in vitro* by chemical carcinogens. Cancer Res., 34, 72-78.

23) CASTO, B.C., PIECZYNSKI, W.J., JANOSKO, N. & DIPAOLO, J.A. (1976). Significance of treatment interval and DNA repair in the enhancement of viral transformation by chemical carcinogens and mutagens. Chem. Bio. Interactions, 13, 105-125.

24) FREEMAM, A.E., PRICE, P.J., IGEL, H.J., YOUNG, J.C., MARYAK, J.M. & HUEBNER, R.J. (1970). Morphological transformation of rat embryo cells induced by diethylnitrosamine and murine leukemia virus. J. Nat. Cancer Inst., 44, 65-78.

25) TRAUL, K.A., TAKAYAMA, K., KACHEVSKY, V., HINK, R.J. & WOLFF, J.S. (1981). A rapid *in vitro* assay for carcinogenicity of chemical substances in mammalian cells utilizing an attachment-independence endpoint. J. Appl. Toxicol., 1, 190-195.

26) ELIS, P.M., YUSPA, S.H., GULLINO, M., MORGAN, D.L., BATES, R.R. & LUTZNER, M.A. (1974). *In vitro* neoplastic transformation of mouse skin cells: Morphology and ultrastructure of cells and tumors. J. Invest. Dermatol., 62, 569-581.

27) YUSPA, S.H., HAWLEY-NELSON, P., KOEHLER, B. & STANLEY, J.R. (1980). A survey of transformation markers in differentiating epidermal cell lines in culture. Cancer Res., 40, 4649-4703.

28) KULESZ-MARTIN, M.F., KOEHLER, B., HENNINGS, H. & YUSPA, S.H. (1980). Quantitative assay for carcinogen altered differenti-

ation in mouse epidermal cells. Carcinogenesis, 1, 995-1006.

29) STEELE, V.E., MARCHOK, A.C. & METTESHEIM, P. (1979). Oncogenic transformation in epithelial cell lines derived from tracheal explants exposed *in vitro* to N-methyl-N'-nitro-N-nitrosoguanidine. Cancer Res., 39, 3805-3811.

30) YANG, N.S., KIRKLAND, W., JORGENSEN, T. & FURMANSKI, P. (1980). Absence of fibronectin and presence of plasminogen activator in both normal and malignant human mammary epithelial cells in culture. J. Cell Biol., 84, 120-130.

31) RICHARDS, J. & NANDI, S. (1978). Neoplastic transformation of rat mammary cells exposed to 7,12-dimethylbenz[a]anthracene or N-nitrosomethylurea in cell culture. Proc. Natl. Aca. Sci. USA, 75, 3836-3840.

32) POILEY, J.A., RAINERI, R. & PIENTA, R.J. (1979). The use of hamster hepatocytes to metabolize carcinogens in an *in vitro* bioassay. J. Natl. Cancer. Inst., 63, 519-524.

33) TU, A., BREEN, P. & SIVAK, A. (1984). Comparison of primary hepatocytes and S-9 metabolic activation systems for the C3H-10T½ cell transformation assay. Carcinogenesis, in press.

34) SIVAK, A. & VAN DUUREN, B.L. (1967). Phenotypic expression of transformation: Induction in cell culture by a phorbol ester. Science, 157, 1443-1444.

35) MONDAL, S., BRANKOW, D.W. & HEIDELBERGER, C. (1976). Two-stage chemical oncogenesis in cultures of C3H-10T½ cells. Cancer Res., 36, 2254-2260.

36) FRAZELLE, J.H., ABERNETHY, D.J. & BOREIKO, C.J. (1983). Factors influencing the promotion of transformation in chemically-initiated C3H/10T½ Cl 8 mouse embryo fibroblasts. Carcinogenesis, 4, 709-715.

37) DORMAN, B.H., BUTTERWORTH, B.E. & BOREIKO, C.J. (1983). Role of intercellular communication in the promotion of C3H/10T½ cell transformation. Carcinogenesis, 4, 1109-1115.

38) SIVAK, A. & TU, A.S. (1980). Cell culture tumor promotion
     experiments with saccharin, phorbol myristate acetate and
     several common food materials. Cancer Lett., 10, 27-32.

39) TRAUL, K.A., HINK, R.J., Jr., KACHEVSKY, V. & WOLFF, J.S.III.
     (1981). Two-stage carcinogenesis *in vitro*: Transformation of
     3-methylcholanthrene-initiated Rauscher murine leukemia virus-
     infected rat embryo cells by diverse tumor promoters. J. Natl.
     Cancer Inst., 66(1), 171-176.

40) POILEY, J.A., RAINERI, R. & PIENTA, R.J. (1979). Two-stage
     malignant transformation in hamster embryo cell. Br. J. Can-
     cer, 39, 8-14.

41) RIVEDAL, E. & SANNER, T. (1981). Metal salts as promoters of
     in vitro morphological transformation of hamster embryo cells
     initiated by benzo(a)pyrene. Cancer Res., 41, 2950-2953.

42) POPESCU, C.N., AMSBAUGH, S.C. & DIPAOLO, J.A. (1980). Enhance-
     ment of N-methyl-N'-nitro-N-nitrosoguanidine transformation
     of Syrian hamster cells by a phorbol diester is independent
     of sister chromatid exchanges and chromosome aberrations.
     Proc. Natl. Acad. Sci. USA, 77, 7282-7286.

43) KAKUNAGA, T. & CROW, J.E. (1973). Cell variants showing dif-
     ferential susceptibility to ultraviolet light-induced trans-
     formation. Science, 209, 505-507.

44) KENNEDY, A.R., FOX, M., MURPHY, G. & LITTLE, J.B. (1980).
     Relationship between X-ray exposure and malignant transform-
     ation in C3H-10T½ cells. Proc. Natl. Acad. Sci. USA, 77, 7262-
     7266.

45) CHAN, G.L. & LITTLE, J.B. (1982). Dissociated occurrence of
     single-gene mutation and oncogenic transformation in C3H-10T½
     cells exposed to ultraviolet light and caffeine. J. Cell.
     Physiol., 111, 309-314.

46) KENNEDY, A.R., CAIRNS, J. & LITTLE, J.B. (1984). Timing of
     the steps in transformation of C3H-10T½ cells by X-irradi-
     ation. Nature, 307, 85-86.

47) RUBIN, H. & CHU, B.M. (1982). Self-normalization of highly transformed 3T3 cells through maximized contact interaction. Proc. Natl. Acad. Sci. USA, 79, 1903-1907.

48) RUBIN, H. & CHU, B.M. (1984). Solute concentration effects on the expression of cellular heterogeneity of anchorage-independent growth among spontaneously transformed BALB/c-3T3 cells. *In Vitro*, 20, 585-596.

49) RUBIN, H. (1984). Adaptive changes in spontaneously transformed BALB/3T3 cells during tumor formation and subsequent cultivation. J. Nat. Cancer Inst., 2, 375-381.

50) RUBIN, H., ARNSTEIN, P. & CHU, B.M. (1984). High-frequency variation and population drift in a newly transformed clone of BALB/3T3 cells. Cancer Res., 44, 5242-5248.

51) NEWBOLD, R.F., OVERELL, R.W. & CONNELL, J.R. (1982). Induction of immortality is an early event in malignant transformation of mammalian cells by carcinogens. Nature, 299, 633-635.

52) NEWBOLD, R.F. & OVERELL, R.W. (1983). Fibroblast immortality is a prerequisite for transformation by EJ c-Ha-ras oncogene. Nature, 304, 648-651.

Supported by U.S. Public Health Service Contract N01-ES-15794.

ARE ACTIVATED PROTO-*ONC* GENES CANCER GENES?*

Peter H. Duesberg[1], *Michael Nunn*[2], *Nancy Kan*[3],
Dennis Watson[3], *Peter H. Seeburg*[4] & *Takis Papas*[3]

[1]Department of Molecular Biology, University of
California, Berkeley, California 94720, USA
[2]The Salk Institute, P.O. Box 85800, San Diego
California 92138-9216, USA
[3]Laboratory of Molecular Oncology, National Cancer
Institute, Frederick Cancer Research Facility
Frederick, Maryland 21701, USA
[4]Genentech, Inc., 460 Point San Bruno Boulevard
South San Francisco, California 90007, USA

ABSTRACT

Cellular genes, which are related to retroviral transforming (*onc*)
genes have, therefore, been termed proto-*onc* genes, are now widely
believed to be potential cancer genes. In some tumors, proto-*onc*
genes are mutated or expressed more than in normal cells. Under
these conditions, proto-*onc* genes are thought to be activated to
function as cancer genes in view of two hypotheses: The one gene-
one cancer hypothesis which suggests that one activated proto-*onc*
gene, like a viral *onc* gene, is sufficient to cause cancer and the
multigene-one cancer hypothesis which speculates that an activated
proto-*onc* gene, unlike a viral *onc* gene, is a necessary but not a

*This lecture was also presented at the "International Conference
on RNA Tumor Viruses in Human Cancer", Denver, Colorado, USA, June
10-14, 1984, and at the sixth meeting on "Modern Trends in Human
Leukemia" at Wilsede, Germany, June 17-21, 1984. A portion of this
lecture will also be printed as part of a review in <u>Science</u>.

sufficient cause of cancer. The evidence for these hypotheses is reviewed here using as examples proto-*myc* and proto-*ras*, the cellular prototypes of the *onc* genes of avian carcinoma virus MC29 and murine Harvey sarcoma virus. Since mutated or transcriptionally activated proto-*onc* genes are not consistently associated with a specific tumor and do not transform primary cells and since as yet no set of an activated proto-*onc* gene and a complementary cancer gene with transforming function has been isolated from a tumor, there is no proof that activated proto-*onc* genes are sufficient or even necessary to cause cancer.

INTRODUCTION

The main objective of cancer molecular biology is to identify cancer genes. Despite fierce efforts, this objective has not yet been met (1-3). Nevertheless, it has been proposed, recently, that molecularly defined or cloned DNA species from some tumors are singular cancer genes, because these DNAs are capable of transforming the morphology of certain preneoplastic cell lines (4). Despite the popularity of this view, there is no convincing evidence to date that these DNA species can also transform normal cells in culture or that they are the causes of tumors in animals (see below). Circumstantial evidence suggests that most cancers are not caused by single genes but are the products of multiple genes that have been formally divided into initiation and promotion or maintenance genes (1-3). Retroviruses without *onc* genes (chronic leukemia viruses) and DNA viruses are thought to function either as initiation or as maintenance genes in multigene cancers because these viruses enhance the cancer risk of infected animals. To date, it is still unknown how these viruses might enhance the cancer risk and which set of viral and cellular genes are needed for a given cancer.

As yet the only known cancer genes are the transforming *onc* genes
of retroviruses. However, carcinogenesis by retroviruses with *onc*
genes does not fit the multigene hypothesis. Typically these viruses
initiate and maintain cancers with singular transforming genes that
are dominant in susceptible cells (5). The discovery of single-gene
determinants of cancer in retroviruses has become a precedent that
has infected cancer gene research. It has made retroviral *onc* genes
the favorite models of cellular oncogenes, although the relevance
of single-gene models to virus-negative tumors is as yet unknown.
Fortunately, *onc* genes are either detrimental or at least useless
to the viability of the virus and thus are not maintained by retroviruses. They are the product of rare, genetic accidents, generated
by illegitimate recombinations between retroviruses and cellular
genes, termed proto-*onc* genes. About twenty different proto-*onc*
genes corresponding to 20 different retroviral *onc* genes are known
(5). At this time, the normal function of proto-*onc* genes has not
yet been determined. One of them is structurally related to a
growth factor, another to a growth factor receptor (6) and two
appear to be yeast cell cycle genes (6,7).

It is now widely believed that, upon transcriptional or mutational
"activation", proto-*onc* genes function like viral *onc* genes. Activation is assumed to be the conversion of a non-oncogenic proto-
onc gene into a carcinogenic variant. Indeed, mutationally altered
or transcriptionally activated proto-*onc* genes have been found in
certain tumors. However, the known mutationally or transcriptionally altered proto-*onc* genes are structurally different from viral
onc genes and have not been shown to be the causes of tumors. There
is as yet no adequate functional evidence for oncogenicity and no
consistent correlation between any proto-*onc* alteration and a certain tumor. To date, viral *onc* genes are the only proven examples
of "activated" proto-*onc* genes.

RETROVIRAL *ONC* GENES AND PROTO-*ONC* GENES

Retroviruses with *onc* genes are the fastest acting, obligatory carcinogens known to date. Such viruses have only been isolated from animals with neoplasms, while all other retroviruses and all DNA viruses with oncogenic potential are regularly isolated from animals without neoplasms. This is consistent with single gene carcinogenesis by retroviruses with *onc* genes and possible multigene carcinogenesis with all other viruses. Indeed, retroviral *onc* genes are the only genes known that initiate and maintain cancers *per se*. That they are necessary for transformation has been proven genetically with temperature-sensitive (*ts*) mutants of Rous (RSV) (8), Kirsten (KiSV) (9), and Fujinami sarcoma viruses (10,11), with avian erythroblastosis virus (12), and with deletion mutants of these and other retroviruses (13-19). The most convincing argument, that they are also sufficient to initiate and maintain neoplastic transformation, is that all susceptible cells infected by retroviruses with *onc* genes become transformed as soon as they are infected. This high transformation efficiency virtually excludes selection of preneoplastic cells initiated by another gene.

The structural characteristic of retroviral *onc* genes is a specific sequence that is unrelated to the three essential virion genes *gag*, *pol* and *env*. This *onc*-specific sequence of retroviruses is related to one or several proto-*onc* genes. Typically, the *onc*-specific sequence replaces essential virion genes and thus renders the virus replication-defective, or it is added to the essential genes as in the case of RSV and is readily deleted (5,13,14,20). Since *onc* sequences are parasitic and have no survival value for the virus, *onc* genes are readily lost by spontaneous deletion (5,20). Therefore, viruses with *onc* genes are subject to extinction unless maintained in laboratories.

About 17 of the 20 known viral *onc* genes are hybrids of coding regions from proto-*onc* genes linked to coding regions from essential
retroviral genes (20). The remaining viral *onc* genes consists of
coding regions from proto-*onc* genes linked to retroviral control
elements. The identification of hybrid *onc* genes provided the first
unambiguous clues that viral *onc* genes and corresponding cellular
proto-*onc* genes are different, since proto-*onc* genes are neither
related to nor linked in the cell to elements of essential retrovirus genes (21,22). Sequence comparisons of cloned genes have
since confirmed and extended that viral *onc* genes and corresponding
proto-*onc* genes are not isogenic (5,20). The known viral *onc* genes
are subsets of proto-*onc* genes linked to regulatory and coding elements of virion genes.

In our laboratories we are studying the structural and functional
relationships between viral *onc* genes and corresponding proto-*onc*
genes, with particular emphasis on the *onc* genes of the following
avian carcinoma, sarcoma and leukemia viruses. The *onc* gene of
avian carcinoma virus MC29 was the first among viral *onc* genes to
be diagnosed as a hybrid gene (21,23) (Fig. 1). About one-half of
its information (1.5 kb) is derived from the *gag* gene of retroviruses; the otherhalf (1.6 kb), termed *myc* is derived from theproto-
myc gene (22). The gene is defined by a 110,000 dalton Δ*gag-myc*
protein, termed p110 (21,24). The proto-*myc* gene of the chicken has
at least 3 exons. The boundaries of the first exon are as yet un-
defined (25-27). The *myc* region of MC29 derives four codons possibly from the 3' end of the first exon and includes the second and
third proto-*myc* exons (Fig. 1). Three other avian carcinoma viruses
MH2, OK10 and CMII also have *onc* genes with *myc* sequences (24). The
myc-related gene of MH2 is derived from the second and third proto-
myc exon and includes the splice acceptor of the first proto-*myc*
intron (25,28,29) (Fig. 1). It also appears to be a hybrid consisting of six *gag* codons up to the splice donor of the *gag* gene (30).

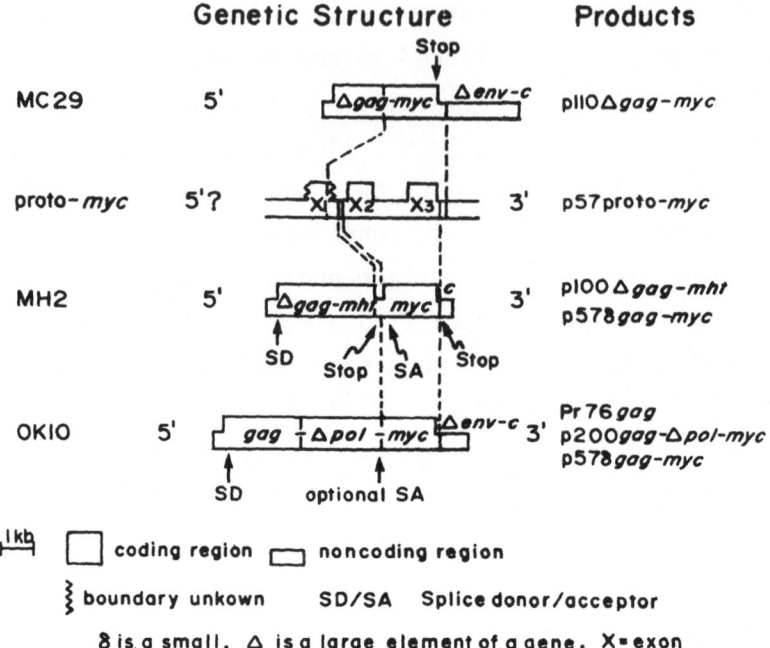

Figure 1. *Comparison of the genetic structures and gene products of the myc-related genes of MC29, MH2, OK10 and chicken proto-myc.*

It is expressed via a subgenomic mRNA as a p57 *myc*-related protein product (31-33). In addition, MH2 contains a second potential transforming gene, Δ*gag-mht*. The *mht* sequence is very closely related to the *onc* gene of murine sarcoma virus MSV 3611 (28,29). It is as yet unclear whether both genes are necessary for transforming function. The *myc* sequence of OK10, like that of MH2, is derived from the second and third proto-*myc* exons and includes the splice acceptor of the first proto-*myc* intron (Fig. 1).(J. Hayflick, P. Seeburg, R. Ohlsson, S. Pfeifer, D. Watson, T. Papas & P. Duesberg, unpublished). It is expressed via a subgenomic mRNA as a p57 protein (32-34). At the same time, the *myc* sequence of OK10 is also part of a large hybrid *onc* gene, *gag-*Δ*pol-myc*, similar to the hybrid *myc* gene of MC29 (24). This gene is defined by a 200,000 dalton protein termed p200 (24). Again, it remains to be determined whether both of these two *onc* gene products are necessary for transforming function. The *myc* sequence of CMII is part of a Δ*gag-myc* hybrid gene similar to that of MC29 (24). Thus, all *myc*-related viral *onc* genes are subsets of proto-*onc* linked to large or small retroviral coding regions and regulatory elements. As yet, no virus with a *myc*-related *onc* gene has been isolated from a mammalian species. However, a *myc*-containing feline provirus with unknown biological activity was recently detected by hybridization of lymphoma DNA from a feline leukemia virus-infected cat (35).

The results of similar comparisons between the Δ*gag-fps* genes of Fujinami, PRCII and PRCIIp sarcoma viruses and cellular proto-*fps* are summarized in Figure 2 (19,36,37). In these cases, the sarcoma viruses share with proto-*fps* a 2- to 3-kb *fps* domain, including probably the 3' translation stop codon. However, the viral genes each initiate with retroviral *fps* regions, whereas proto-*fps* initiates with a proto-*pfs*-specific exon(s) (36) (Fig. 2).

Analysis of the *onc* genes of the leukemia viruses avian myeloblastosis (AMV) and erythroblastosis virus (E26), and of proto-*myb*, the common cellular prototype of the *myb* sequence shared by these viruses, are also schematically summarized in Figure 2. Unlike the *myb*- and *fps*-containing *onc* genes, the *onc* genes of each of these viruses share an internal domain with the cellular prototype (38, 39). In E26, the *myb* region is flanked by a *gag*-related region at its 5' and by a newly discovered *onc*-specific domain, termed *ets*, at its 3' end to form a tripartite *onc* gene (40,41). In AMV, the *myb* region includes a proto-*myb* splice acceptor that is presumably served in the virus by the splice donor of Δ*gag* (30). The *myb* region of AMV is flanked at its 3' end by an element derived from the *env* gene of retroviruses. It is concluded that the *onc*-specific sequences of each of these carcinoma, sarcoma and leukemia viruses

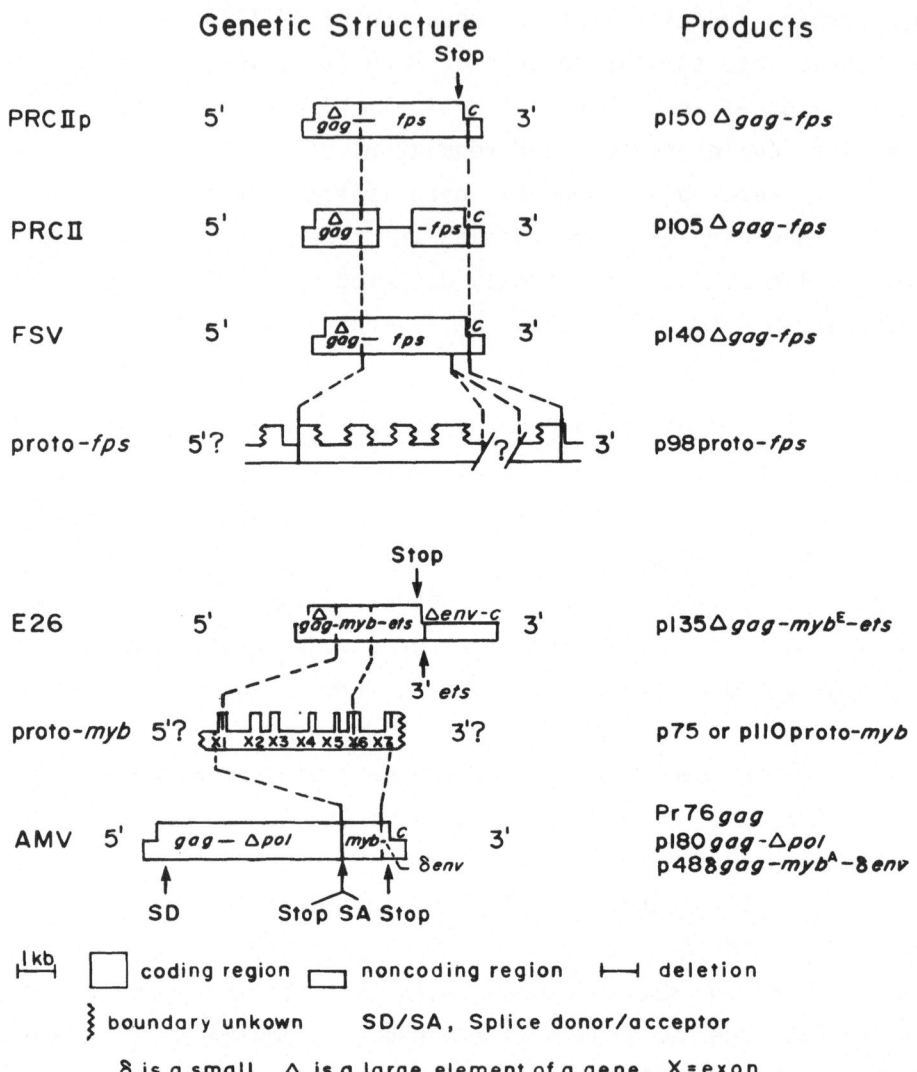

Figure 2. *Comparison of the genetic structures and gene products of the fps-related genes of avian Fujinami, PRCIIp and PRCII sarcoma viruses and the chicken proto-fps gene (top) and of the myb-related genes of avian leukemia viruses E26 and AMV and the chicken proto-myb gene (bottom).*

are subsets of proto-*onc* genes linked to elements of essential
retrovirus genes.

Other examples of hybrid *onc* genes have been described (5,16,20,
24). Since in all cases studied, proto-*onc* genes are not related
and not linked to essential genes of retroviruses, all viral
hybrid *onc* genes are by definition structurally different from
proto-*onc* genes. The coding regions of a few viral *onc* genes, like
the *src* gene of RSV and probably the *onc* genes of Harvey and Kirsten sarcoma viruses (termed Ha- and Ki-*ras*) are derived entirely
from proto-*onc* sequences (see below and Fig. 4). Nevertheless,
even these *onc* genes differ from proto-*onc* genes in extensive
deletions and point mutations. For example, the *src* gene of RSV is
a hybrid of genetic elements derived from at least three proto-*src*
genes (5,42).

Two arguments indicate that these qualitative differences between
onc and proto-*onc* genes are essential for transforming function of
the viral genes: There is the overwhelming evidence that many proto-
onc genes are regularly expressed in normal cells without altering
the normal phenotype (5,43). There is more indirect evidence that
proto-*onc* sequences cloned in retroviral or plasmid vectors do not
transform normal, diploid cells. For example, phage or plasmid
vectors carrying the viral *src*-related region, but not a complete
complement of the major proto-*src* gene (44-47) or proto-*fos* , the
precursor of the transforming gene of FBJ murine osteosarcoma virus
(48), or proto-*fps*/*fes*, the precursors of avian Fujinami and feline
sarcoma viruses (49) (W.-H. Lee & P. Duesberg, unpublished), or
proto-*myc*, the precursor of avian MC29 virus (T. Robins, P. Duesberg & G. Vande Woude, unpublished), do not transform cells in culture. The *src*-related region of the major proto-*src* gene also fails
to transform in a RSV vector (50). Further, proto-*src* and proto-Ha-
ras (the precursor of Ha-MuSV) fail to transform in a reticuloendo-

theliosis virus vector while the corresponding viral *onc* genes have transforming function (51).

Apparent exceptions are proto-*mos* and proto-*ras* which, after ligation to retroviral promoters, transform the preneoplastic NIH 3T3 cell line (52,53). The proto-*mos* and *ras* regions used in these constructions are essentially the same as those found in Moloney and Harvey sarcoma viruses but are not complete proto-*onc* genes (see below and Fig. 2). Conceivably, the proto-*onc* regions that were not included into these constructions and are not in the viruses, might in the cell suppress transforming potential of the complete proto-*onc* genes. Moreover, it will be detailed below that transforming function in 3T3 cells is not a reliable measure of transforming function in diploid embryo cells or in the animal. Neither the proto-*ras* nor the proto-*mos* construction were found to transform diploid embryo cells (54,55) (and G. Vande Woude, personal communication). Thus, normal proto-*onc* genes and viral *onc* genes are related, but are structurally and functionally different. The question is now whether there are conditions under which proto-*onc* genes can cause cancer.

THE SEARCH FOR ACTIVATION OF PROTO-*ONC* GENES TO CHANGE CANCER GENES

The only clear, although indirect, proof for activation of proto-*onc* genes to cancer genes is based on the rare cases in which proto-*onc* genes functioned as accidental parents of retroviral *onc* genes. It has been deduced from structural analyses of retroviral genes and proto-*onc* genes that viral *onc* genes were generated by transduction of specific domains from proto-*onc* genes (5,20). Because no significant sequence homology exists between retroviruses and proto-*onc* genes, such transductions must procede via two rare, non-homologous recombinations (5,25). In addition, it appears that only a few cellular genes are proto-*onc* genes or can function as progenitors or viral *onc* genes since the same proto-*onc* sequences have been found in different viral isolates (29). It is probably for this reason that viral transductions or "activations" are extremely rare, even though all cells contain proto-*onc* genes and many animal species contain retrovirus without *onc* genes. Only 50 to 100 sporadic cancers from which retroviruses with *onc* genes were isolated have been reported and no experimentally reproducible system of transduction has ever been described (56-58). Thus, retroviruses with *onc* genes are the causes of rare, natural tumors, rather than laboratory artifacts.

Their role as accidental progenitors of viral *onc* genes has made proto-*onc* genes the focus of the search for cellular cancer genes. Their possible function in cancer was initially tested in many laboratories in view of a "one gene-one cancer" and more recently in view of a "multigene-one cancer" hypothesis. The one gene-one cancer hypothesis, essentially the oncogene hypothesis of Huebner and Todaro, postulates that activation of inactive cellular oncogenes is sufficient to cause cancer (59). Some investigators have postulated that activation is the result of increased dosage of a given proto-*onc* gene product. This view, termed the quantitative model, received support from early experiments which suggested that the *src* gene of RSV or the *myc* gene of MC29 and the corresponding proto-*onc* genes were equivalent (60-64). In the meantime, significant structural and functional differences between these genes have been found (5,43,44-47,50) (see above). Others have suggested that proto-*onc* genes are activated by mutations or rearrangements in the primary DNA sequence (65,66). This view is termed the qualitative model (5).

The multigene-one cancer hypothesis postulates that an activated proto-*onc* gene is necessary, but unlike the corresponding viral gene, not sufficient to cause cancer. A quantitatively or quali-

tatively activated proto-*onc* gene is postulated to function either
as initiation or as maintenance gene together with another proto-
onc gene, in a multistep process (54,55,67-73). This hypothesis fits
the view of how virus-negative tumors are thought to arise in general and provides identifiable candidates to test the hypothesis.
However, since retroviral *onc* genes have yet to be dissociated into
initiation and maintenance functions, this hypothesis is without
viral precedent.

Two kinds of assays have been performed to test these hypotheses.
One assay correlates transcriptional activation and mutation of
proto-*onc* genes with cancer; the other directly measures transforming function of proto-*onc* genes upon transfection into certain
recipient cells, typically the preneoplastic mouse NIH 3T3 cell
line (4,54,55). Such experiments have most frequently linked cancers with alterations of proto-*myc* and proto-*ras*.

Is proto-*myc* activation the cause of B-cell lymphomas?

Based on the observation that transcription of the cellular proto-
myc is enhanced in retroviral lymphomas of chicken, it has been
postulated that transcriptional activation of proto-*myc* is the
cause of B-cell lymphoma (64,74). Chicken B-cell lymphoma is a
clonal cancer that appears in a small franction of animals infected
by one of the avian leukosis viruses (which have no *onc* genes)
after latent periods of over six months (58). The hypothesis, termed downstream promotion, postulates that the gene is activated by
the promoter of a retrovirus integrated upstream (Fig. 3) and that
activated proto-*myc* functions like the transforming gene of MC29
(64). Subsequently, samples were found in which the retrovirus is
integrated 3' of proto-*myc* or 5' in the opposite transcriptional
direction. In these cases, the virus is thought to function like
an enhancer of proto-*myc* (74) (Fig. 3).

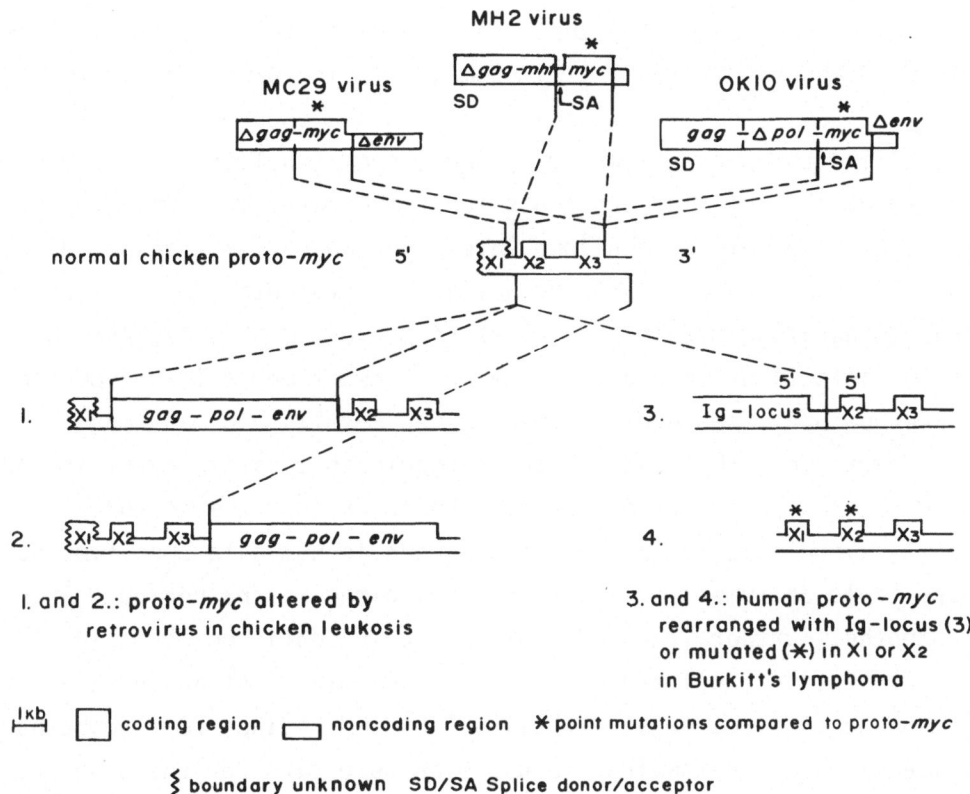

Figure 3.   *Myc-related genes in avian carcinoma viruses and in normal and lymphoma cells.* The common and specific *myc* domains
of avian carcinoma viruses MC29 (25,26), MH2 (28,29) and
OK10 (Ref. 24 and unpublished), of normal chicken proto-
myc (25,71), and of the proto-*myc* genes of avian leukosis (57,67) and human Burkitt's lymphoma (70,71,80) are
graphically compared. Proto-*myc* has three exons (X1, X2,
X3) the first of which is thought to be noncoding (25,
81). The proto-*myc* genes of chicken and man are related
but not identical: Their first exons are essentially
unrelated, there are major unique sequence elements in
each of their second exons and minor differences between
the third exons (25). *Gag, pol, env* are the three essential virion genes of retroviruses and Δ marks incomplete
complements of these genes.

However, proto-*myc* differs structurally from the 3-kb Δ*gag-myc* gene
of MC29 as diagrammed in Figures 1 and 3 (25,26). Further, it has
been argued previously (5) that the hypothesis fails to explain the
origin of about 20% of viral lymphomas in which proto-*myc* is not
activated (64); the discrepancies between the phenotype of the
disease and the cancers caused by MC29; the clonality of the tumors,
defined by a single integration site of the retrovirus with regard
to proto-*myc* as well as the long latent period of the disease. Given
about 10^6 kb of chicken DNA and activation of proto-*myc* by retrovirus integration within about 5 kb of proto-*myc* (27,74), one in
2×10^5 infections should generate the first tumor cell. Since the
chicken probably has over 10^7 uncommited B-cells and many more
virus particles, the critical carcinogenic integration event should
occur after a short latent period. The tumor should also not be
clonal, since integration by retroviruses is not site-specific and
there could be numerous infections during the latent period of about
six months. Further, the model has not been confirmed in murine
(75,76), feline (35), and bovine (77) leukemia. Instead the high
percentage of virus-negative feline (35) and bovine (78) lymphomas
indicates that a retrovirus is not even necessary for the disease.

Recently, it was suggested that a mutation, rather than a virus,
may have activated avian proto-*myc* because mutations have been
observed in viral lymphoma (79). However, the proto-*myc* mutations
have not been shown to be the cause of the viral lymphoma.

Activation of proto-*myc* has also been postulated to cause the
retrovirus-negative, human Burkitt's lymphomas, and mouse plasma-
cytomas. In these cases, chromosome translocation has been proposed
as a mechanism of activating proto-*myc* function (70,71,80,81). The
human proto-*myc* is related to that of the chicken from which carcinoma viruses have been derived (Fig. 3). The two genes have unique
first exons, similar second exons with unique regions and colinear

third exons (25). In man, proto-*myc* is located on chromosome 8 and
an element of this chromosome is reciprocally translocated in many
Burkitt's lymphoma lines to immunoglobulin (Ig) loci of chromosome
14 and less frequently of chromosome 2 or 22. Since the crossover
points of chromosome 8 are near proto-*myc*, translocation was initially suspected to activate proto-*myc* transcriptionally by re-
arranging proto-*myc* (Fig. 1) or by altering its immediate environment and thus bringing it under the influence of new promoters or
enhancers (80). However, in many lymphomas rearranged proto-*myc* is
not linked to a new promoter, instead the first presumably noncoding exon is replaced by the Ig locus, linked to it 5'-5' in the
opposite transcriptional orientation (80) (Fig. 3). Further this
model cannot explain how proto-*myc* would be activated when the
complete proto-*myc* gene, including its known promoters and flanking
regions, is translocated (70,72,82), or recent observations that in
a significant minority of Burkitt's lymphomas proto-*myc* remains in
its original chromosomal location while a region 3' of proto-*myc*
is translocated (83-87). Despite these inconsistencies, proto-*myc*
is thought to function as a cellular oncogene in these tumors.

Moreover, there is no consensus at this time whether proto-*myc* expression is enhanced in Burkitt's lymphoma cells, as compared to
normal control cells. Some investigators report elevated expression
compared to normal B-lymphoblasts or lines (88), while others report essentially normal levels of proto-*myc* mRNA (70,82,86,87,89-
92). Further, enhanced proto-*myc* transcription is not specific for
B-cell lymphomas, since high levels of proto-*myc* expression are
seen in non-Burkitt's lymphomas (91), in other tumors (73), and in
chemically transformed fibroblast cell lines in which proto-*myc* is
not translocated or rearranged (43). The view that enhanced expression of proto-*myc* may be sufficient to cause Burkitt's lymphoma
is also challenged by the observation that proto-*myc* transcription
reaches either cell cycle-dependent peak levels in certain cell

lines (43,93) or maintains constitutively high levels in embryo
cells similar to those in tumor cells (F. Cuzin (Nice), M. Bywater
(Uppsala) & A. Braithwaite (Canberra), personal communications).

The possibility that mutations of proto-*myc* may correlate with
Burkitt's lymphoma has also been investigated. In some Burkitt's
cell lines mutations have been observed in translocated, but un-
arranged, proto-*myc* (93,94) (Fig. 3). Initially, it was proposed
that these mutations may activate proto-*myc* by altering the gene
product (94), but in at least one Burkitt's lymphoma line the coding sequence corresponding to proto-*myc* exons 2 and 3 was identical
to that of the normal gene (82) (Fig. 3). Recently, it has been
proposed that mutations in the first noncoding exon may activate
the gene (92,95). However, there is no functional evidence for this
view and an activating mutation that is characteristic of Burkitt's
lymphomas has not been identified. It is also an open question at
this time whether the first human proto-*myc* exon is indeed noncoding (82) or has possibly a large, open reading frame capable of
encoding a major protein (25,95a). A sequence comparison between
translocated proto-*myc* of a mouse plasmacytoma with the germline
proto-*myc* found the two genes to be identical except for one nucleotide difference in the first exon. It was concluded that proto-*myc*
mutations are not required for oncogenes (96).

Therefore, no translocation, rearrangement, elevated expression,
or characteristic mutation of proto-*myc* is common to all Burkitt's
lymphomas investigated. This casts doubt on the concept that any
of the known proto-*myc* alterations are a sufficient cause (or even
necessary) for Burkitt's lymphoma.

The question of whether proto-*myc* has transforming function has
been tested directly using the 3T3 cell-transformation assay with
DNA from chicken or human B-cell lymphomas. However, no *myc*-related

DNA was detected even though its presumed functional equivalent, the Δgag-myc gene of MC29, is capable of transforming 3T3 cells (97,98) and other rodent cell lines (99). Instead, another DNA sequence, termed Blym was identified by the assay (67,100). Based on these results, the role of proto-myc in lymphomas has been interpreted in terms of two-gene hypothesis. It has been suggested that activated proto-myc is necessary but not sufficient to cause the lymphoma (68,70). It is postulated to have a transient early function that generates a lymphoma maintenance gene, Blym. This gene appears to be the DNA that transforms 3T3 cells and is thought to maintain the B-cell tumor. There is no proof for this postulated role of proto-myc as a lymphoma initiation gene, because the 3T3 cell-transformation assay does not measure proto-myc initiation function, and because there is no evidence that the two genes jointly (or alone) transform B-cells. Furthermore, the hypothesis does not address the question why proto-myc should have any trans-forming function at all, if it is not like MC29. (MC29 does not require a second gene to transform a susceptible cell.) It is also not known whether Blym is altered in primary Burkitt's lymphomas, since all of the transfection experiments were done with DNA from cell lines.

It is conceivable that chromosome translocation involving the proto-myc chromosome 8 may be a specific but not a necessary consequence, rather than the cause of the lymphoma (101). Human B-cell lymphomas with translocations that do not involved chromosome 8 have indeed been described (102,103). In the case of clonal myeloid leukemias with consistent translocations, like the Philadelphia chromosome, it has been convincingly argued that translocation is preceded by clonal proliferation of certain stem cells with the same isoenzyme markers as leukemic cells but without chromosomal or clinical ab-normalities (104).

Perhaps primary Burkitt's lymphomas should be analysed now and more emphasis should be given to the question whether proto-*myc* alteration contributes to Burkitt's lymphoma, rather than to speculation about possible mechanisms.

Proto-*ras* mutations, the cause of human and rodent carcinomas?

Use of the 3T3 cell assay to measure transforming function of DNA from a human bladder carcinoma cell line has identified DNA homologous to the *ras* gene of Harvey sarcoma virus (66,104) (Fig. 4). Based on the viral model, the proto-Ha-*ras* gene is thought to be a potential cancer gene because it encodes a 21,000-dalton protein, p21, which is colinear with an *ras* gene product p21 of Ha-MuSV (106) (Fig. 4). The proto-Ha-*ras* gene from the bladder carcinoma cell line differs from normal proto-Ha-*ras* in a point mutation which alters the 12th p21 codon in exon 1 from normal *gly* to *val* (66, 107). This mutation does not cause overproduction of the *ras* gene product (p21) in the 3T3 cell line (66) and does not change known biochemical properties of p21 (108). The single base change is thought to activate the gene to a functional equivalent of Ha-MuSV and to be the cause of the carcinoma because it is the apparent cause for 3T3 cell-transforming function (66,109). However, this mutation has not been found in over 60 primary human carcinomas, including 10 bladder, 9 colon and 10 lung carcinomas (110), in 8 other lung carcinomas (111), and 14 additional bladder and 9 kidney carcinomas (R. Muschel & G. Khoury, personal communication). Further the mutated human proto-Ha-*ras*, which transforms 3T3 cells, does not transform primary rat embryo cells (54,69) and, more significantly, does not transform human embryo cell (112). Transformation of primary cells would be expected from a gene that causes tumors in animals. Thus the mutated proto-*ras* gene does not correspond to the viral model which transforms primary mouse, rat (113,114) and human cells (115-119). In addition, the *val* in the

12th codon of 3T3 cell transforming proto-*ras* is different for the
the *arg* of the viral counterpart (107).

Other mutations have since been found to confer 3T3 cell-transforming function to proto-Ha-*ras* DNA. Proto-Ha-*ras* with a mutation in
codon 61 was isolated from a human tumor cell line (120). 3T3 cell-
transforming proto-Ha-*ras* DNAs were also isolated from 2 out of 23
primary urinary tract tumors analyzed. One of these contained a
mutation in codon 61, the other was not identified (121). The mutations were not found in the normal tissue of the respective patients. Nevertheless, this does not prove that 3T3 cell-transforming function of proto-*ras* was necessary for tumor formation since
each was associated with only 1 out of 23 histologically indistinguishable tumors.

A 3T3 cell-transforming mouse proto-Ha-*ras* DNA was also found in
some (not all) chemically induced benign papillomas and malignant
carcinomas of mice (122). Since only a small (5-7%) portion of the
benign tumors progressed to carcinomas, it would appear that 3T3
cell-transforming proto-*ras* was not sufficient to cause the carcinomas, and since not all carcinomas contained the mutation, it
would appear that it was not necessary either. A high proportion,
i.e. 14 out of 17 methylnitrosourea-induced mammary carcinomas of
rats, were found to contain 3T3 cell-transforming proto-Ha-*ras* DNA
(M. Barbacid, personal communication). This suggests that the mutation is not necessary for the tumor, although it may be important
for tumor progression. The original study reported 9 out of 9 posi-
tives (123). Moreover, the hormone-dependence and high tissue specificity of the carcinogen in this study suggests that other genes
must be involved, because mutated proto-*ras* has been found in association with other tumors and transforms 3T3 cells without hormones.
It is plausible that other genes, which may be involved in tumorigenesis but which do not register in the 3T3 assay, were also
altered by the carcinogen.

In an effort to explain why mutated proto-Ha-*ras* transforms pre-neoplastic 3T3 cells, but not rat or human embryo cells, it has recently been proposed that mutated proto-Ha-*ras* is only one of at least two activated genes that are necessary to induce cancer (54, 55,69). This two-gene hypothesis has been tested by transfecting primary rat cells with a mixture of the mutated human proto-Ha-*ras* and either MC29 provirus or activated proto-*myc* from mouse plasma-cytoma (54), or the EIA gene of adenovirus (69) as helper genes. None of these genes were able to transform rat embryo cells by themselves, but some cells were transformed by the artificial mixed doubles. The study that used the adenovirus virus helper gene show-ed that proto-*ras* expression varied from high to normal levels in transformed cells and that normal proto-*ras* was inactive in the assay (69). The study that used *myc*-related helper genes did not show that the transformants contained and expressed the added DNAs. It also did not test whether unaltered forms of proto-*myc* or proto-*ras* were sufficient for a mixture of these genes to register in this assay. This appears to be a particularly relevant question since a proto-*myc* clone from a mouse plasmacytoma with an SV40 en-hancer at its 3' end but without its natural promoter (71) was re-ported to be active (54) although such a construction is not ex-pected to activate proto-*myc*.

The *myc*-related genes were proposed to convert rat embryo cells to cells that are capable of dividing indefinitely, like 3T3 cells, a function termed immortalization (54,55). The supposed immortaliz-ation function of MC29 or of activated proto-*myc* was not demon-strated independently. The proposal did not explain why an immor-talization gene was necessary. Obviously immortalization is necess-ary to maintain cells in culture. However, immortalization is not necessary for focus formation and probably not for tumor formation since embryo cells are capable of sufficient rounds of mitoses (up to 50) in cell culture and in the animal (124). In the avian sys-

tem, MC29 transforms primary cells and causes tumors in chicken
independently without the benefit of secondary oncogenes, and most
MC29 tumor cells are not immortal if tested in cell culture. The
failure of maintaining cells from many human tumors in cell culture,
under conditions where cells from similar tumors survive, also suggests that immortality may not be an essential criterion of a tumor
cell (125, 125a). There is also no precedent for a function of proto-*ras*
in a multistep transformation mechanism, because the transforming
genes of Harvey or Kirsten sarcoma viruses transform rat and mouse
embryo cells (113,114) or human embryo cells (115-119) with single
hit kinetics and without helper genes. Moreover, there is no precedent for the artifical mixtures of the two activated proto-*onc*
genes in any natural tumors.

Other 3T3 cell-transforming proto-*ras* genes, namely proto-Ki-*ras*,
which is more closely related to the *ras* gene of Kirsten sarcoma
virus than to Harvey virus, and N-*ras*, which is related to both
viruses, have also been found in tumors or cell lines (126). Proto-
Ki-*ras* encodes a p21 protein that is related to the p21 protein
encoded by proto-Ha-*ras* (107,126,127). One group has found 3T3
cell-transforming proto-Ki-*ras* DNA in three primary human tumors
and five tumor cell lines out of 96 samples tested (111,128). The
same group also found 3T3 cell-transforming proto-Ki-*ras* DNA in
one out of eight lung carcinomas tested (111). The DNA from this
tumor, but not that from normal tissue of the same patient, had a
mutation in the 12th codon. Obviously the low percentage of 3T3
cell-positives among these tumors rasises the question of whether
the mutations were necessary for tumorigenesis.

In a study of human melanomas, only one of five different meta-
stases from the same human melanoma patient was found to contain
3T3 cell-transforming proto-Ki-*ras* DNA (129). A 3T3 cell-transforming Ki-*ras* DNA was also detected in a metastatic variant but not

in a primary methylcholanthrene-induced T-cell lymphoma of mice
(130). An example of a spontaneous proto-*ras* mutation appearing in
tumor cells cultured *in vitro* has just been described (131). This
suggests that these proto-*ras* mutations were consequences rather
than the causes of these tumors. The view that *ras* mutation is a
consequence of tumorigenesis is also consistent with the results
that only one *ras* allele is mutated in some primary tumors (111,
121, 127) whereas both alleles are mutated in typical tumor cell
lines (110,111).

Since 3T3 cell-transforming or mutated proto-*ras* genes are only
rarely associated with human and murine tumors and since mutated
proto-Ha-*ras* does not transform human or rat embryo cells (54,69,
112) (proto-Ki-*ras* was not tested), there is as yet no proof that
mutated proto-*ras* is sufficient or even necessary for any of the
above tumors.

The failure of the mutated proto-Ha- or Ki-*ras* to behave like the
viral model suggests that structural differences between the cellular and viral genes are responsible (Fig. 4). The 5' end of proto-
Ha-*ras* is not as yet defined (107). Proto-Ha-*ras* differs from the
5.5 kb RNA genome of Harvey sarcoma virus (132) in a cell-specific
1 kb DNA region 5' of exon 1 that is preceded by a virus-related
region (107) and in the sizes (1.2 and 5 kb) of the proto-*ras*
transcripts compared to the genomic viral 5.5 kb mRNA (58,133,134).
The cell-specific proto-Ha-*ras* region is thought to be an intron
but it may have another function. The base changes that confer 3T3
cell-transforming function to proto-Ha-*ras* are different from those
that set apart viral *ras* genes from proto-*ras* (66,107,126) (Fig. 4).
Proto-Ha-*ras* with 3T3 cell-transforming function further differs
from the viral *ras* and from normal proto-Ha-*ras* in point mutations
in exons 1 or 2 (66,107) (Fig. 4). Moreover, only about 10% of the
genomes of Harvey and Kirsten sarcoma viruses, are *ras*-related.

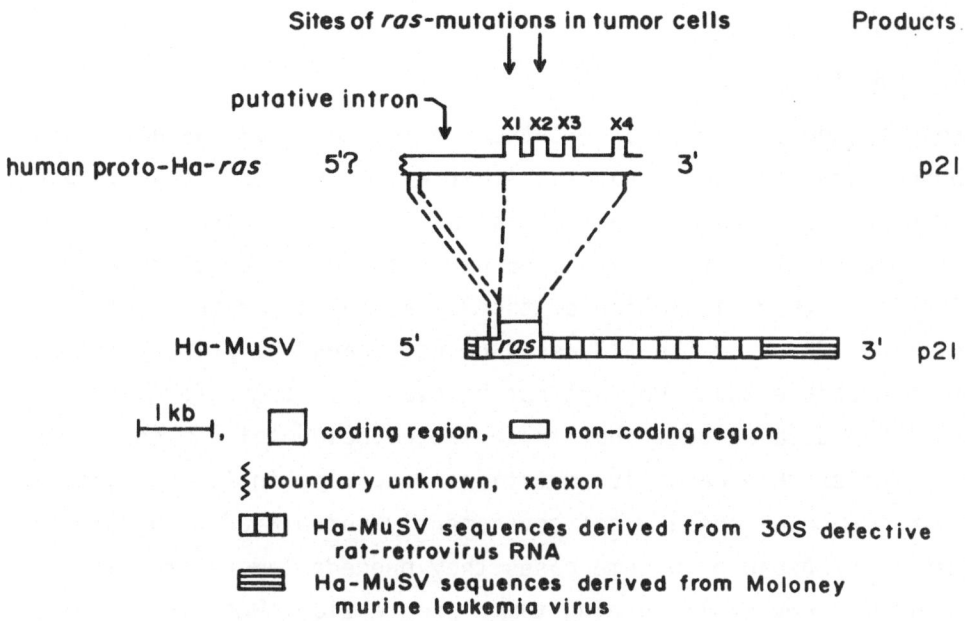

Figure 4. *Comparison of the genetic structures and p21 gene prod-
           ucts of the human proto-Ha-ras gene* (106,107) *and the
           5.5 kb RNA genome of Harvey sarcoma virus (Ha-MuSV)*
           (132). Ha-MuSV is a genetic hybrid of the rat proto-*ras*
           gene, a 30S defective retrovirus RNA from rat cells and
           of Moloney leukemia virus (107,135).

Each viral RNA contains about 3 kb of genetic information, derived
from a rat 30S defective retrovirus RNA (135) which may contribute
to the oncogenicity of these viruses (Fig. 4). Further, it has been
argued that mutated proto-*ras* is a recessive transforming gene,
because both *ras* alleles are mutated in typical tumor cell lines,
although only one allele is mutated in some primary tumors (111,
127). By contrast, the viral *ras* gene is dominant. A definitive
answer to the question whether *ras* mutations are dominant or recessive 3T3 cell-transforming genes could be obtained by simultaneous transformation with mutated and normal *ras* genes. Finally,
Ha- and Ki-MuSV are not obvious models for proto-*ras* genes with
hypothetical carcinoma function, since these viruses cause pre-
dominantly sarcomas.

CONCLUSIONS

Does the 3T3 assay detect cancer genes?

The preponderance of 3T3 cell-transformation negatives among the
above described tumors suggests that either no genes have caused
the negative tumors or that the assay failed to detect them. That
only *ras*-related proto-*onc* genes have been detected in human tumors
signals another limitation of the 3T3 assay. Since the proto-*ras*
mutations found by the 3T3 assay do not transform primary cells,
it is possible that they are not relevant for tumor formation.
Available data suggest that these are coincidental or consequen-
tial rather than causative mutations occurring in tumor cells, because the mutations are not consistently correlated with specific
tumors and because in some cases they precede tumor formation and
in others they evolve during tumor progression. Despite its effectiveness to transform 3T3 cells, it would follow that mutated proto-*ras*
is not a dominant singular cancer gene, similar to a viral *onc* gene, and
that the test is insufficient to determine whether proto-*onc* genes cause
tumors in animals. The efficiency of the assay to identify cancer genes
unrelated to proto-*onc* genes (4) remains to be determind.

Are altered proto-*onc* genes sufficient causes of cancer?

Clearly, proto-*onc* genes are sometimes mutationally or transcriptionally altered in tumor cells. However, no altered proto-*onc*
gene has been found that looks like a viral *onc* gene. More importantly, no altered proto-*onc* gene from tumors investigated functions
like a viral *onc* gene. Altered proto-*myc* has no transforming function in known assay systems, and altered proto-*ras* transforms 3T3
cells but does not transform rodent or human embryo cells. Thus,
altered proto-*onc* genes are structurally and functionally different from viral *onc* genes. Moreover, altered proto-*onc* genes are not
consistently associated with specific tumors. Since there is no
functional evidence that altered proto-*onc* genes transform embryo
cells or cause tumors and no consistent correlation between altered

proto-*onc* genes and a specific tumor, the one-gene hypothesis (that
altered proto-*onc* genes are sufficient to cause tumors) is without
support. As yet, viral *onc* genes are the only "activated" proto-
genes that are sufficient to cause tumors.

Are altered proto-*onc* genes necessary to cause cancer?

The observations that altered proto-*onc* genes do not behave like
viral *onc* genes and that in some tumors multiple proto-*onc* genes
are altered (73) have been interpreted in terms of a multigene
hypothesis. Altered proto-*myc* has been proposed to cooperate with
the *Blym* gene to cause chicken and human B-cell lymphoma (68).
Altered proto-*ras* has been proposed to cause carcinomas with other
genes, and reported to cooperate in an artificial system with
altered proto-*myc* to transform rat embryo cells in culture (54,55).
However, there are several reservations about a role of altered
proto-*myc* or proto-*ras* in multigene carcinogenesis: (i) There is
no functional evidence that a combination of altered *myc* and *Blym*
from lymphomas or that altered *ras*, together with another gene
from carcinomas, transform appropriate normal test cells. An artificial combination of altered *ras* in combination with a *myc*-related
or an adenovirus gene was reported to transform primary rat cells.
However, it was not reported whether both genes are present and
functional in all transformants, and there is no evidence that,
these artificial *ras*-helper genes are models for the hypothetical
helper genes in tumors with altered *ras*. (ii) The observations that
proto-*myc* alterations are not consistently associated with B-cell
lymphomas and that proto-*ras* mutations are only rarely associated
with specific carcinomas argue that at least one of two hypothetic-
ally synergistic cancer genes is not necessary for these tumors.
As yet, no multigene complements that include one or two proto-*onc*
genes have been shown to be consistently associated with specific
tumors. (iii) The proposals that altered proto-*onc* genes play a
role in a multigene carcinogenesis are a significant departure

from the original view that they were equivalents of viral *onc*
genes. The proposals speculate that altered proto-*onc* genes are
necessary but not sufficient for tumor formation and behave like
functional subsets of viral *onc* genes. They do not address the
question why these genes are assumed to have unique oncogenic functions that are different from those of the viral models. The *ad hoc*
assumption is without precedent since it is not known whether viral
onc genes can be dissociated into complementary or helper gene-
dependent genetic subsets. Since there is no functional proof for
multiple, synergistic transforming genes and no consistent correlation between at least one altered proto-*onc* gene and a specific
tumor, the view that proto-*onc* genes are necessary for multigene
carcinogenesis is speculative.

Prospects:

It may be argued that the proto-*onc* gene alterations that are associated with some cancers play a nonspecific but causative role
in carcinogenesis that could be substituted for by another gene.
To support this view, it would be necessary to know which other
genes could substitute for the role that altered proto-*onc* genes
are thought to play in the origin of cancer. Further, one would
have to know whether proto-*onc* gene alterations are more typical
of cancer cells that alterations of other genes and which other
genes characteristically undergo alterations in tumor cells. It is
likely that unknown events, additional to the known alterations of
resident proto-*onc* genes, are required for the development of cancer (5,136).

The fact that proto-*onc* genes share common domains with viral
genes remains a persuasive argument that proto-*onc* genes may, under
certain conditions, be changed into cancer genes. The evidence that
most normal proto-*onc* genes are expressed in normal celle suggests
that cell-specific domains of proto-*onc* genes may suppress poten-

tial oncogenic function. Thus, mutation or removal of suppressors could activate a proto-*onc* gene, as has been predicted for Burkitt's lymphoma. Clearly, the identification of such suppressors would depend on the complete genetic and biochemical definition of proto-*onc* genes. To date, we do not know both termini of any proto-*onc* (except for human proto-*myc* (82) which is not a prototype of a known oncogenic virus). The addition, virus specific *onc* gene elements may also be essential to activate a proto-*onc* gene. In this case, a retrovirus without an *onc* gene (chronic leukemia virus) could activate a proto-*onc* gene by a single illegitimate recombination which would form a hybrid *onc* gene. Such an event would be more probable that the generation of a retrovirus with an *onc* gene for which at least two illegitimate recombinations are necessary.

It is remarkable that DNA technology has made it possible to convert nontransforming DNA from viral or cellular sources to DNA species that transform cell lines or embryo cells. Examples are the proto-*mos* and proto-*ras* retroviral LTR recombinants that transform 3T3 cells (52,53), the proto-*ras* , *myc* and adenovirus DNA combinations that transform rat embryo cells (54,69), or an LTR-mutant proto-*ras*-SV40 construction that transforms rat embryo cells (137). Another example is a synthetic gene that consists of a mouse proto-*myc* gene in which all or part of the first exon is replaced by the LTR of mouse mammary tumor virus. Upon introduction into the germ line, this gene was expressed in 11 transgenic mice. Two of these developed mammary tumors after two pregnancies, but not in all mammary glands. It was suggested that the gene may be necessary but not sufficient for the development of these tumors (138). Both the level of expression and the integrity of proto-*onc* genes were altered in these constructions, since only subsets of proto-*onc* genes were included. In order to assess the relevance of such iatrogenic transformations to cancer, it would be helpful to determine whether the number of DNA species that can be converted to transforming variants is large or small, and it would be necessary to determine whether any such DNAs ever occur in natural tumors. The most important challenge now is to develop functional assays for cellular cancer genes.

ACKNOWLEDGEMENTS

I thank my colleagues M. Botchan, M. Carey, G.S. Martin, H. Rubin, W. Phares, S. Pfaff & C. Romerdahl for encouragement and many critical comments and L. Brownstein for typing numerous drafts of this manuscript. The work from my laboratory is supported by NIH grant Ca 11426 from the National Cancer Institute and by grant CTR 1547 from The Council for Tobacco Research - U.S.A., Inc.

REFERENCES

1)   ROUS, P. (1967). The challenge to man of the neoplastic cell. Science, 157, 24.

1a) BERENBLUM, I. (1981). Sequential aspects of chemical carcino-genesis: Skin. In: "Cancer, Vol. 1", F. Becker, ed., Plenum Press, New York.

2)   KNUDSON, A.G. Genetic influences in human tumors, *ibid.*

3)   FOULDS, L. (1969). "Neoplastic Development, Vol. I & II", Academic Press, New York.

4)   COOPER, G.M. (1982). Cellular transforming genes, Science, 218, 801.

5)   DUESBERG, P.H. (1983). Retroviral transforming genes in normal cells?, Nature, 304, 219.

6)   HELDIN, C.-H. & WESTERMARK, B. (1984). Growth factors: Mechanism of action and relation to oncogenes, Cell, 37, 9.

7)   PETERSON, T.A., YOCHEM, J., BYERS, B., NUNN, M.F., DUESBERG, P.H., DOOLITTLE, R.F. & REED, S.E. (1984). A relationship between the yeast cell cycle genes *CDC4* and *CDC36* and the *ets* sequence of oncogenic virus E26, Nature, 309, 556.

8)   MARTIN, G.S. (1970). Rous sarcoma virus: A function required
     for the maintenance of the transformed state, Nature, 221, 1021.

9)   SHIH, T.Y., WEEKS, M.O., YOUNG, M.A. & SCOLNICK, E.M. (1979).
     p21 of Kirsten murine sarcoma virus is thermolabile in a viral
     mutant temperature sensitive for the maintenance of transform-
     ation, J. Virol., 31, 546.

10) PAWSON, A., GUYDEN, J., KUNG, T.-H., RADKE, K., GILMORE, T. &
     MARTIN, G.S. (1980). A strain of Fujinami sarcoma virus which
     is temperature-sensitive in protein phosphorylation and cellu-
     lar transformation, Cell, 22, 767.

11) LEE, W.-H., BISTER, K., MOSCOVICI, C. & DUESBERG, P.H. (1981).
     Temperature-sensitive mutants of Fujinami sarcoma virus:
     Tumorigenicity and reversible phosphorylation of the trans-
     forming p140 protein, J. Virol., 38, 1064.

12) PALMIERI, S., BEUG, H. & GRAF, T. (1982). Isolation and char-
     acterization of four new temperature-sensitive mutants of
     avian erythroblastosis virus (AEV), Virology, 123, 296.

13) DUESBERG, P.H. & VOGT, P.K. (1970). Differences between the
     ribonucleic acids of transforming and nontransforming avian
     tumor viruses, Proc. Natl. Acad. Sci. USA, 67, 1673.

14) MARTIN, G.S. & DUESBERG, P.H. (1972). The a-subunit in the RNA
     of transforming avian tumor viruses: I. Occurrence in differ-
     ent virus strains. II. Spontaneous loss resulting in nontrans-
     forming variants, Virology, 47, 494.

15) WEI, C.-M., LOWY, D.R. & SCOLNICK, E.M. (1980). Mapping of
     transforming region of the Harvey murine sarcoma virus genome
     by using insertion-deletion mutants constructed *in vitro*,
     Proc. Natl. Acad. Sci. USA, 77, 4674.

16) GOFF, S.P. & BALTIMORE, D. (1982). The cellular oncogene of
     the Abelson murine leukemia virus genome. In: "Advances in
     Viral Oncology, Volume 1", G. Klein, ed., Raven Press, New
     York.

17) SRINIVASAN, A., DUNN, C.Y., YUASA, Y., DEVARE, S.G., REDDY, E.P. & AARONSON, S.A. (1982). Abelson murine leukemia virus: Structural requirements for transforming gene function, Proc. Natl. Acad. Sci. USA, 79, 5508.

18) EVANS, L.H. & DUESBERG, P.H. (1982). Isolation of a transform- ation-defective deletion mutant of Moloney murine sarcoma virus, J. Virol., 41, 735.

19) DUESBERG, P.H., PHARES, W. & LEE, W.H. (1983). The low tumori- genic potential pf PRCII, among viruses of the Fujinami sar- coma virus subgroup, corresponds to an internal *fps* deletion of the transforming gene, Virology, 131, 144.

20) DUESBERG, P.H. (1980). Transforming genes of retroviruses, Cold Spring Harbor Symp. Quant. Biol., 44, 13.

21) MELLON, P., PAWSON, A., BISTER, K., MARTIN, G.S. & DUESBERG, P.H. (1978). Specific RNA sequences and gene products of MC29 avian acute leukemia virus, Proc. Natl. Acad. Sci. USA, 75, 5874.

22) ROBINS, T., BISTER, K., GARON, C., PAPAS, T. & DUESBERG, P.H. (1982). Structural relationship between a normal chicken DNA locus and the transforming gene of the avian acute leukemia virus MC29, J. Virol., 41, 635.

23) DUESBERG, P.H., BISTER, K. & VOGT, P.K. (1977). The RNA of avian acute leukemia virus MC29, Proc. Natl. Acad. Sci. USA, 74, 4320.

24) BISTER, K. & DUESBERG, P.H. (1982). Genetic structure and transforming genes of avian retroviruses. In: "Advances in Viral Oncology, Volume 1", G. Klein, ed., Raven Press, New York.

25) PAPAS, T.S., KAN, N.K., WATSON, D.K., FLORDELLIS, C.S., PSALLIDOPOULOS, M.C., LAUTENBERGER, J., SAMUEL, K.P. & DUESBERG, P.H. (1982). *myc*-related genes in viruses and cells. In: "Cancer Cells 2/Oncogenes and Viral Genes", G.F. Wande Woude, A.J. Levine, W.C. Topp & J.D. Watson, eds., Cold Spring

Harbor Laboratory, Cold Spring Harbor, New York.

26) WATSON, D.K., REDDY, E.P., DUESBERG, P.H., PAPAS, T.S. (1983). Nucleotide sequence analysis of the chicken c-*myc* gene reveals homologous and unique coding regions by comparison with the transforming gene of avian myelocytomatosis virus MC29 Δ*gag-myc*, Proc. Natl. Acad. Sci. USA, 80, 2146.

27) SHIH, C.-K., LINIAL, M., GOODENOW, M.M. & HAYWARD, W.S. (1984). Nucleotide sequence 5' of the chicken c-*myc* coding region: Localization of a noncoding exon that is absent from *myc* transcripts in most avian leukosis virus-induced lymphomas, Proc. Natl. Acad. Sci. USA, 81, 4697.

28) KAN, N.C., FLORDELLIS, C.S., MARK, G.E., DUESBERG, P.H. & PAPAS, T.S. (1984). Nucleotide sequence of avian carcinoma virus MH2: Two potential *onc* genes, one related to avian virus MC29, the other to murine sarcoma virus 3611, Proc. Natl. Acad. Sci. USA, 81, 3000.

29) KAN, N.C., FLORDELLIS, C.S., MARK, G.E., DUESBERG, P.H. & PAPAS, T.S. (1984). A common *onc* gene sequence transduced by avian carcinoma virus MH2 and by murine sarcoma virus 3611, Science, 223, 813.

30) SCHWARTZ, D.E., TIZARD, R. & GILBERT, W. (1983). Nucleotide sequence of Rous sarcoma virus, Cell, 32, 853.

31) PACHL, C., BIEGALKE, B. & LINIAL, M. (1983). RNA and protein encoded by MH2 virus: Evidence for subgenomic expression of v-*myc*, J. Virol. 45, 133.

32) HANN, S.R., ABRAMS, H.D., ROHRSCHNEIDER, L.R. & EISENMAN, R.N. (1983). Proteins encoded by v-*myc* and c-*myc* oncogenes: Identification and localization in acute leukemia virus transformants and bursal lymphoma in cell lines, Cell, 34, 781.

33) ALITALO, K., RAMSAY, G., BISHOP, J.M., PFEIFFER, S.O., COLBY, W.W. & LEVINSON, A.D. (1983). Identification of nuclear proteins encoded by viral and cellular *myc* oncogenes, Nature, 306, 274.

34) CHISWELL, D.J., RAMSEY, G. & HAYMAN, M.J. (1981). Two virus-
     specific RNA species are present in cells transformed by de-
     fective leukemia virus OK10, J. Virol., 40, 301.

35) LEVY, L.S., GARDNER, M.B. & CASEY, J.W. (1984). Isolation of
     a feline leukaemia provirus containing the oncogene myc from
     a feline lymphosarcoma, Nature, 308, 853.

36) SEEBURG, P.H., LEE, W.-H., NUNN, M.F. & DUESBERG, P.H. (1984).
     The 5' end of the transforming gene of Fujinami sarcoma virus
     and of the cellular proto-fps gene are not colinear, Virology,
     133, 460.

37) LEE, W.-H., PHARES, W. & DUESBERG, P.H. (1983). Structural
     relationship between chicken DNA locus, proto-fps, and the
     transforming gene of Fujinami sarcoma virus (Δgag-fps),
     Virology, 129, 79.

38) RUSHLOW, K.E., LAUTENBERGER, J.A., PAPAS, T.S., BALUDA, M.A.,
     PERBAL, B., CHIRIKJIAN, J.G. & REDDY, E.P. (1982). Nucleotide
     sequence of the transforming gene of avian myeloblastosis
     virus, Science, 216, 1421.

39) KLEMPNAUER, K.-H., GONDA, T.S. & BISHOP, J.M. (1982). Nucleo-
     tide sequence of the retroviral leukemia gene v-myb and its
     progeintor c-myb: The architecture of a transduced oncogene,
     Cell, 31, 453.

40) NUNN, M.F., SEEBURG, P.H., MOSCOVICI, C. & DUESBERG, P.H.
     (1983). Tripartite structure of the avian erythroblastosis
     virus E26 transforming gene, Nature, 306, 391.

41) NUNN, M., WEIHER, H., BULLOCK, P. & DUESBERG, P.H. (1984).
     Avian erythroblastosis virus E26: Nucleotide sequence of the
     tripartite onc gene and of the LTR, and analysis of the cellu-
     lar prototype of the viral ets sequence, Virology, 139, 330.

42) TAKEYA, T. & HANAFUSA, H. (1983). Structure and sequence of
     the cellular gene homologous to the src gene of RSV and the
     mechanism of the generation of the viral transforming gene,
     Cell, 32, 881.

43) CAMPISI, J., GRAY, H.E., PARDEE, A.B., DEAN, M. & SONENSHEIN,
     G.E. (1984). Cell-cycle control of c-*myc* but not c-*ras* express-
     ion is lost following chemical transformation, Cell, 36, 241.

44) TAKEYA, T. & HANAFUSA, H. (1982). DNA sequence of the viral
     and cellular *src* gene of chickens II. Comparison of the *src*
     genes of two strains of avian sarcoma virus and of the cellu-
     lar homolog, J. Virol., 44, 12.

45) PARKER, R.C., VARMUS, H.E. & BISHOP, J.M. (1984). Expression
     of v-*src* and chicken c-*src* in rat cells demonstrates qualita-
     tive differences between pp60^{v-src} and pp60^{c-src}, Cell, 37,
     131.

46) PARSONS, J.T., BRYANT, D., WILKERSON, V., GILMARTIN, G. &
     PARSONS, S.J. (1984). Site-directed mutagenesis of Rous sarcoma
     virus pp60^{src}: Identification of functional domains required
     for transformation. In: "Cancer Cells 2/Oncogenes and Viral
     Genes", G.F. Vande Woude, A.J. Levine, W.C. Topp & J.D. Watson,
     eds., Cold Spring Harbor Laboratory, Cold Spring Harbor, New
     York.

47) SHALLOWAY, D., COUSSENS, P.M. & YACIUK, P.   (1984). c-*src* and
     src homolog overexpression in mouse cells, *ibid*.

48) MILLER, A.D., CURRAN, T. & VERMA, I.M. (1984). c-*fos* protein
     can induce cellular transformation: A novel mechanism of
     activation of a cellular oncogene, Cell, 36, 51.

49) SODROSKI, J.G., GOH, W.C. & HASELTINE, W.A. (1984). Transform-
     ing potential of a human protooncogene (c-*fps/fes*) locus,
     Proc. Natl. Acad. Sci. USA, 81, 3039.

50) IBA, H., TAKEYA, T., CROSS, F.R., HANAFUSA, T. & HANAFUSA, H.
     (1984). Rous sarcoma virus variants that carry the cellular
     src gene instead of the viral *src* gene cannot transform
     chicken embryo fibroblasts, Proc. Natl. Acad. Sci. USA, 81,
     4424.

51) WILHELMSEN, K.C., TARPLEY, W.G. & TEMIN, H.M. (1984). Identi-
     fication of some of the parameters governing transformation

by oncogenes in retroviruses. In: "Cancer Cells 2/Oncogenes and Viral Genes", G.F. Wande Woude, A.J. Levine, W.C. Topp & J.D. Watson, eds., Cold Spring Harbor Laboratory, Cold Spring Harbor, New York

52) BLAIR, D.G., OSKARSSON, M., WOOD, T.G., McCLEMENTS, W.C., FISCHINGER, P.J. & VANDE WOUDE, G.F. (1981). Activation of the transforming potential of a normal cell sequence: A molecular model for oncogenesis, Science, 212, 941.

53) CHANG, E.H., FURTH, M.E., SCOLNICK, E.M. & LOWY, D.R. (1982). Tumorigenic transformation of mammalian cells induced by a normal human gene homologous to the oncogene of Harvey murine sarcoma virus, Nature, 297, 479.

54) LAND, H., PARADA, L.F. & WEINBERG, R.A., (1983). Tumorigenic conversion of primary embryo fibroblasts requires at least two cooperating oncogenes, Nature, 304, 596.

55) LAND, H., PARADA, L.F. & WEINBERG, R.A. (1983). Cellular oncogenes and multistep carcinogenesis, Science, 222, 771.

56) GROSS, L. (1970). "Oncogenic Viruses", Pergamon Press, New York.

57) TOOZE, J., ed. (1973). "The Molecular Biology of Tumour Viruses", Cold Spring Harbor Laboratory, Cold Spring Harbor, New York.

58) WEISS, R.A., TEICH, N.M., VARMUS, H. & COFFIN, J.M., eds., (1982). "Molecular Biology of Tumor Viruses: RNA Tumor Viruses, Cold Spring Harbor Laboratory, Cold Spring Harbor, New York.

59) HUEBNER, R.J. & TODARO, G.J. (1969). Oncogenes of RNA tumor viruses as determinants of cancer, Proc. Natl. Acad. Sci. USA, 64, 1087.

60) BISHOP, J.M., COURTNEIDGE, S.A., LEVINSON, A.D. OPPERMANN, H., QUINTRELL, N., SHEINESS, D.K., WEISS, S.R. & VARMUS, H.E. (1980). Origin and function of avian retrovirus transforming genes, Cold Spring Harbor Symp. Quant. Biol., 44, 919.

61) BISHOP, J.M. (1981). Enemies within: The genesis of retrovirus oncogenes, Cell, 23, 5.

62) WANG, L.-H., SNYDER, P., HANAFUSA, T., MOSCOVICI, C. & HANAFUSA, H. (1980). Comparative analysis of cellular and viral sequences related to sarcomagenic cell transformation, Cold Spring Harbor Symp. Quant. Biol., 44, 766.

63) KARESS, R.E., HAYWARD, W.S. & HANAFUSA, H. (1980). Transforming proteins encoded by the cellular information of recovered avian sarcoma viruses, Cold Spring Harbor Symp. Quant. Biol., 44, 765.

64) HAYWARD, W.S., NEEL, B.G. & ASTRIN, S.M. (1981). Activation of a cellular *onc* gene by promoter insertion in ALV-induced lymphoid leukosis, Nature, 290, 475.

65) KLEIN, G. (1981). The role of gene dosage and genetic transpositions in carcinogenesis, Nature, 294, 313.

66) TABIN, C.J., BRADLEY, S.M., BARGMANN, C.I., WEINBERG, R.A., PAPAGEORGE, A.G., SCOLNICK, E.M., DHAR, R., LOWY, D.R. & CHANG, E.H. (1982). Mechanism of activation of a human oncogene, Nature, 300, 143.

67) COOPER, G.M. & NEIMAN, P.E. (1981). Two distinct candidate transforming genes of lymphoid leukosis virus-induced neoplasms, Nature, 292, 857.

68) DIAMOND, A., COOPER, G.M., RITZ, J. & LANE, M.-A. (1983). Identification and molecular cloning of the human *Blym* transforming gene activated in Burkitt's lymphomas, Nature, 305, 112.

69) RULEY, H.E. (1983). Adenovirus early region 1A enables viral and cellular transforming genes to transform primary cells in culture, Nature, 304, 602.

70) LEDER, P., BATTEY, J., LENOIR, G., MOUDLING, C., MURPHY, W., POTTER, H., STEWART, T. & TAUB, R. (1983). Translocations among antibody genes in human cancer, Science, 222, 765.

71) ADAMS, J.M., GERONDAKIS, S., WEBB, E., CARCORAN, L.M. & CORY, S. (1983). Cellular *myc* oncogene is altered by chromosome translocation to an immunoglobulin locus in murine plasma-cytoma and is rearranged similarly in human Burkitt lymphomas, Proc. Natl. Acad. Sci. USA, 80, 1982.

72) KLEIN, G. & KLEIN, E. (1984). Oncogene activation and tumor progression, Carcinogenesis, 5, 429.

73) SLAMON, D.J., DeKERNION, J.B., VERMA, I.M. & CLINE, M.J. (1984). Expression of cellular oncogenes in human malignancies, Science, 224, 256.

74) PAYNE, G.S., BISHOP, J.M. & VARMUS, H.E. (1982). Multiple arrangements of viral DNA and an activated host oncogene in bursal lymphomas, Nature, 295, 209.

75) TSICHLIS, P.N., STRAUSS, P.G. & HU, L.F. (1983). A common region for proviral DNA integration in MoMuLV-induced rat thymic lymphomas, Nature, 302, 445.

76) YOSHIMURA, F.K. & LEVINE, K.L. (1983). AKR thymic lymphomas involving mink cell focus-inducing leukemia viruses have a common region of provirus integration, J. Virol., 45, 576.

77) KETTMANN, R., DESCHAMPS, J., CLEUTER, Y., COUEZ, D., BURNY, A. & MARBAIX, G. (1982). Leukemogenesis by bovine leukemia virus: Proviral DNA integration and lack of RNA expression of viral long terminal repeat and 3' proximate cellular sequences, Proc. Natl. Acad. Sci. USA, 79, 2465.

78) MILLER, J.M., MILLER, L.D., OLSON, C. & GILLETTE, K.S. (1969). Virus-like particles in phytohemagglutinin-stimulated lympho-cyte cultures with reference to bovine lymphosarcoma, J. Natl. Cancer Inst., 43, 1297.

79) WESTAWAY, D., PAYNE, G. & VARMUS, H.E. (1984). Proviral dele-tions and oncogene base-substitutions in insertionally muta-genized c-*myc* alleles may contribute to the progression of avian bursal tumors, Proc. Natl. Acad. Sci. USA, 81, 843.

80) KLEIN, G. (1983). Specific chromosomal translocations and the

genesis of B-cell-derived tumors in mice and men, Cell, 32, 311.

81) ROWLEY, J.D. (1983). Human oncogene locations and chromosome aberrations, Nature, 301, 290.

82) BATTEY, J., MOULDING, C., TAUB, R., MURPHY, W., STEWART, T., POTTER, H., LENOIR, G. & LEDER, P. (1983). The human c-*myc* oncogene: Structural consequences of translocation into the IgH locus in Burkitt lymphoma, Cell, 34, 779.

83) GELMANN, E., PSALLIDOPOULOS, M.C.,PAPAS, T.S. & DALLA-FAVERA, R.(1983). Identification of reciprocal translocation sites within the c-*myc* oncogene amd immunoglobulin μ locus in a Burkitt lymphoma, Nature, 306, 799.

84) CROCE, C.M., THIERFELDER, W., ERIKSON, J., NISHIKURA, K., FINAN, J., LENOIR, G.M. & NOWELL, P.C. (1983). Transcriptional activation of an unrearranged and untranslocated c-*myc* oncogene by translocation of a C_λ locus in Burkitt lymphoma cells, Proc. Natl. Acad. Sci. USA, 80, 6922.

85) ERIKSON, J., AR-RUSHIDI, A., DRWINGA, H.L., NOWELL, P.C. & CROCE, C.M. (1983). Transcriptional activation of the translocated c-*myc* oncogene in Burkitt lymphoma, Proc. Natl. Acad. Sci. USA, 80, 820.

86) HOLLIS, G.F., MITCHELL, K.F., BATTERY, J., POTTER, H., TAUB, R.,LENOIR, G.M. & LEDER, P. (1984). A variant translocation places the λ immunoglobulin genes 3' to the c-*myc* oncogene in Burkitt's lymphoma, Nature, 307, 752.

87) DAVIS, M., MALCOLM, S. & RABBITTS, T.H. (1984). Chromosome translocation can occur on either side of the c-*myc* oncogene in Burkitt lymphoma cells, Nature, 308, 286.

88) ERIKSON, J., NISHIKURA, K., AR-RUSHDI, A., FINAN, J., EMANUEL, B., LENOIR, G., NOWELL, P.C. & CROCE, C.M. (1983). Translocation of an immunoglobulin κ locus to a region 3' of an unrearranged c-*myc* oncogene enhances c-*myc* transcription, Proc. Natl. Acad. Sci. USA, 80, 7581.

89) WESTIN, E.H., WONG-STAAL, F., GELMANN, E.P., DALLA FAVERA, R.
     PAPAS, T.S., LAUTENBERGER, J.A., EVA, A., REDDY, E.P.,
     TRONICK, S.R., AARONSON, S.A. & GALLO, R.C. (1982). Expression
     of cellular homologues of retroviral *onc* genes in human hema-
     topoietic cells, Proc. Natl. Acad. Sci. USA, 79, 2490.

90) MAGUIRE, R.T., ROBINS, T.S., THORGERSSON, S.S. & HEILMAN, C.A.
     (1983). Expression of cellular *myc* and *mos* genes in undiffer-
     entiated B cell lymphomas of Burkitt and non-Burkitt types,
     Proc. Natl. Acad. Sci. USA, 80, 1947.

91) HAMLYN, P.H. & RABBITTS, T.H. (1983). Translocation joins
     c-*myc* and immunoglobulin $_\lambda$1 genes in a Burkitt lymphoma reveal-
     ing a third exon in the c-*myc* oncogene, Nature, 304, 135.

92) TAUB, R., MOULDING, C., BATTEY, J., MURPHY, W., VASICEK, T.,
     LENOIR, G.M. & LEDER, P. (1984). Activation and somatic muta-
     tion of the translocated c-*myc* gene in Burkitt lymphoma cells,
     Cell, 36, 339.

93) KELLY, K., COCHRAN, B.H., STILES, C.D. & LEDER, P.(1983).
     Cell-specific regulation of the c-*myc* gene by lymphocyte mito-
     gens and platelet derived growth factor, Cell, 35, 603.

94) RABBITTS, T.H., HAMLYN, P.H. & BAER, R. (1983). Altered nucleotide
     sequences of a translocated c-*myc* gene in Burkitt lymphoma,
     Nature, 306, 760.

95) RABBITTS, T.H., FORSTER, A., HAMLYN, P. & BAER, R. (19).
     Effect of somatic mutation within translocated c-*myc* genes in
     Burkitt's lymphoma, Nature, 309, 593.

95a) GAZIN, C., DUPONT DE DINECHIN, S., HAMPE, A., MASSON, J.-M.,
     MARTIN, P., STEHELIN, D. & GALIBERT, F. (1984). Nucleotide
     sequence of the human c-*myc* locus: provocative open reading
     frame within the first exon, EMBO J., 3, 383.

96) STANTON, L.W., FAHRLANDER, P.D., TESSER, P.M. & MARCU, K.B.
     (1984). Nucleotide sequence comparison of normal and trans-
     located murine c-*myc* genes, Nature, 310, 423.

96) STANTON, L.W., FAHRLANDER, P.D., TESSER, P.M. & MARCU, K.B.
    (1984). Nucleotide sequence comparison of normal and trans-
    located murine c-*myc* genes, Nature, 310, 423.

97) COPELAND, N.G. & COOPER, G.M. (1980). Transfection by DNAs of
    avian erythroblastosis virus and avian myelocytomatosis virus
    strain MC29, J. Virol., 33, 1199.

98) LAUTENBERGER, J.A., SCHULZ, R.A., GARON, C.F., TSICHLIS, P.H.
    & PAPAS, T.S. (1981). Molecular cloning of avian myeloblasto-
    sis virus (MC29) transforming sequences, Proc. Natl. Acad.
    Sci. USA, 78, 1518.

99) QUADE, K. (1979). Transformation of mammalian cells by avian
    myelocytomatosis virus and avian erythroblastosis virus,
    Virology, 98, 461.

100) GOUBIN, G., GOLDMAN, D.S., LUCE, J., NEIMAN, P.E. & COOPER,
     G.M. (1983). Molecular cloning and nucleotide sequence of a
     transforming gene detected by transfection of chicken B-cell
     lymphoma DNA, Nature, 302, 114.

101) RUBIN, H. (1984). Chromosome aberratons and oncogenes: Cause
     or consequence in cancer, Nature, 309, 518.

102) ERIKSON, J., FINAN, J., TSUJIMOTO, Y., NOWELL, P.C. & CROCE,
     C.(1984). The chromosome 14 breakpoint in neoplastic B cells
     with the t(11;14) translocation involves the immunoglobulin
     heavy chain locus, Proc. Natl. Acad. Sci. USA, 81, 4144.

103) YUNIS, J.J., OKEN, M.D., KAPLAN, M.E., ENSURD, K.M., HOWE, R.R.
     & THEOLOGIDES, A. (1982). Distinctive chromosomal abnormali-
     ties in histologic subtypes of non-Hodgkin's lymphoma, New
     Eng. J. of Med., 307, 1231.

104) FIALKOW, R.J. & SINGER, J.W. (1984). Tracing development and
     cell lineages in human hemopoietic neoplasia. In: "Proceed-
     ings of the Dahlem Workshop on Leukemia", Springer-Verlag,
     Berlin, Germany, in press.

105) DER, J.C., KRONTIRIS, T.G. & COOPER, G.M. (1982). Transform-
     ing genes of human bladder and lung carcinoma cell lines are

homologous to the *ras* genes of Harvey and Kirsten sarcoma
viruses, Proc. Natl. Acad. Sci. USA, 79, 3637.

106) ELLIS, R.W., LOWY, D.R. & SCOLNICK, E.M. (1982). Mouse cells
contain two distinct *ras* gene mRNA species that can be translated into a p21 *onc* protein. In: "Advances in Viral Oncology,
Volume 1", G. Klein, ed., Raven Press, New York.

107) CAPON, D.J., CHEN, E.Y., LEVINSON, A.D., SEEBURG, P.H. &
GOEDDEL, D.V. (1983). Complete nucleotide sequences of the
T24 human bladder carcinoma oncogene and its normal homologue,
Nature, 302, 33.

108) FINKEL, T., CHANNING, J.D. & COOPER, G.M. (1984). Activation
of *ras* genes in human tumors does not affect localization,
modification, or nucleotide binding properties of p21, Cell,
37, 151.

109) REDDY, E.P., REYNOLDS, R.K., SANTOS, E. & BARBACID, M. (1982).
A point mutation is responsible for the acquisition of transforming properties by the T24 human bladder carcinoma oncogene, Nature, 300, 149.

110) FEINBERG, A.P., VOGELSTEIN, B., DROLLER, M.J., BAYLIN, S.B. &
NELKIN, B.D. (1983). Mutation affecting the 12th amino acid
of the c-Ha-*ras* oncogene product occurs infrequently in human
cancer, Science, 220, 1175.

111) SANTOS, E., MARTIN-ZANCA, D., REDDY, E.P., PIEROTTI, M.A.,
DELLA PORTA, G. & BARBACID, M. (1984). Malignant activation
of a K-*ras* oncogene in lung carcinoma but not in normal tissue of the same patient, Science, 223, 661.

112) SAGER, R., TANAKA, K., LAU, C.C., EBINA, Y. & ANISOWICZ, A.
(1983). Resistance of human cells to tumorigenesis induced
by cloned transforming genes, Proc. Natl. Acad. Sci. USA, 80,
7601.

113) HARVEY, J.J. & EAST, J. (1971). The murine sarcoma virus
(MSV), Int. Rev. of Exp. Pathol., 10, 265.

114) LEVY, J.A. (1973). Demonstration of differences in murine sarcoma virus foci formed in mouse and rat cells under a soft agar overlay, J. Nat. Cancer Inst., 46, 1001.

115) AARONSON, S.A. & TODARO, G.I. (1970). Transformation and virus growth by murine sarcoma virus in human cells, Nature, 225, 458.

116) AARONSON, S.A. & WEAVER, C.A. (1971). Characterization of murine sarcoma virus (Kirsten) transformation of mouse and human cells, J. Gen. Virol., 13, 245.

117) KLEMENT, V., FRIEDMAN, M., McALLISTER, R., NELSON-REES, W. & HUEBNER, R.J. (1971). Differences in susceptibility of human cells to mouse sarcoma virus, J. Natl. Cancer Inst., 47, 65.

118) PFEFFER, L.M. & KOPEOLVICH, L. (1977). Differential genetic susceptibility of cultured human skin fibroblasts to transformation of Kirsten murine sarcoma virus, Cell, 10, 313.

119) LEVY, J.A. (1975). Host range of murine xenotropic virus: Replication in avian cells, Nature, 253, 140.

120) YUASA, Y., SRIVASTAVA, S.K., DUNN, C.Y., RHIM, J.S., REDDY, E.P. & AARONSON, S.A. (1983). Acquisition of transforming properties by alternative point mutations within c-*bas*/*has* human proto-oncogene, Nature, 303, 775.

121) FUJITA, J., YOSHIDA, O., YUASA, Y., RHIM, J.S., HATANAKA, M. & AARONSON, S.A. (1984). Ha-*ras* oncogenes are activated by somatic alterations in human urinary tract tumors, Nature, 309, 464.

122) BALMAIN, A., RAMSDEN, M., BOWDEN, G.T. & SMITH, J. (1984). Activation of the mouse cellular Harvey-*ras* gene is chemically induced benign skin papillomas, Nature, 307, 658.

123) SUKUMAR, S., NOTARIO, V., MARTIN-ZANCA, D. & BARBACID, M. (1983). Induction of mammary carcinomas in rats by nitrosomethylurea involves malignant activation of H-*ras*-1 locus by single point mutations, Nature, 306, 658.

124) HOLLIDAY, R. (1983). Cancer and cell senescence, Nature, 306, 742.

125) FOGH, J., ed. (1975). "Human Tumor Cells *In Vitro*", Plenum Press, New York.

125a) SALMON, S.E. (1980). Cloning of human tumor stem cells. Alan R. Liss, N.Y.

126) WIGLER, M., FASANO, O. TAPAROWSKY, E., POWERS, S., KATAOKA, T., BRINBAUM, D., SHIMIZU, K.F. & GOLDFARB, M. (1984). Structure and activation of *ras* genes. In: "Cancer Cells 2/Oncogenes and Viral Genes", G.F. Vande Woude, A.J. Levine, W.C. Topp & J.D. Watson, eds., Cold Spring Harbor Laboratory, Cold Spring Harbor, New York.

127) CAPON, D.J., SEEBURG, P.H., McGRATH, J.P., HAYFLICK, J.S., EDMAN, U., LEVINSON, A.D. & GOEDDEL, D.V. (1983). Activation of Ki-*ras* 2 gene in human colon and lung carcinomas by two different point mutations, Nature, 304, 507.

128) PULCIANI, S., SANTOS, E., LAUVER, A.V., LONG, L.K., AARONSON, S.A. & BARBACID, M. (1982). Oncogenes in solid human tumors, Nature, 300, 539.

129) ALBINO, A.P., LE STRANGE, R., OLIFF, A.I., FURTH, M.E. & OLD, L.J. (1984). Transforming *ras* genes from human melanoma: A manifestation of tumor heterogeneity?, Nature, 308, 69.

130) VOUSDEN, K.M. & MARSHALL, C.J. (1984). Three different activated *ras* genes in mouse tumours; evidence for oncogene activation during progression of a mouse lymphoma, EMBO J., 3, 913.

131) TAINSKY, M.A., COOPER, C.S., GIOVANELLA, B.C. & VANDE WOUDE, G.F. (1984). An activated ras^N gene: Detected in late but not early passage human PA1 teratocarcinoma cells, Science, 225, 643.

132) MAISEL, J., KLEMENT, V., LAI, M.M.C., OSTERTAG, W. & DUESBERG, P.H. (1973). Ribonucleic acid components of murine sarcoma and leukemia viruses, Proc. Natl. Acad. Sci. USA, 70, 3536.

133) ELLIS, R.W., DEFEO, D., FURTH, M.E. & SCOLNICK, E.M. (1982). Mouse cells contain two distinct *ras* gene mRNA species that can be translated into a p21 *onc* protein, Molec. Cell. Biol. 2, 1339.

134) PARADA, L.F., TABIN, C., SHIH, C. & WEINBERG, R.A. (1982). Human EJ bladder carcinoma oncogene is homologue of Harvey sarcoma *ras* gene, Nature, 297, 474.

135) SCOLNICK, E.M., VASS, W.C., HOWK, R.S. & DUESBERG, P.H. (1979). Defective retrovirus-like 30S RNA species of rat and mouse cells are infectious if packaged by Type C helper virus, J. Virol., 29, 964.

136) TEMIN, H.M. (1983). We still don't understand cancer, Nature, 302, 656.

137) SPANDIDOS, D.A. & WILKIE, N.M. (1984). Malignant transformation of early passage rodent cells by a single mutated human oncogene, Nature, 310, 469.

138) STEWART, T.A., PATTENGALE, P.K. & LEDER, P. (1984). Spontaneous mammary adenocarcinoma in transgenic mice that carry and express MTV/*myc* fusion genes, Cell, 38, 627.

IMMUNOGLOBULIN GENES, ONCOGENES, AND HUMAN B-CELL TUMORS

Peter C. Nowell[1] & *Carlo M. Croce*[2]

[1]Department of Pathology and Laboratory of Medicine
University of Pennsylvania School of Medicine
Philadelphia, Pennsylvania 19104, USA

[2]The Wistar Institute of Anatomy and Biology
Philadelphia, Pennsylvania 19104, USA

INTRODUCTION

In the last several years, a combined investigative approach
utilizing techniques of somatic cell genetics, molecular genetics,
and cytogenetics has provided exciting preliminary evidence concerning specific genes important in the pathogenesis of human B-
cell lymphomas and possible mechanisms by which their function may
be altered in neoplastic cells. Burkitt's lymphoma has provided
the most information to date, but other lymphomas and leukemias
are now being explored. Overall, the data are still quite limited,
but the field is moving very rapidly.

In the present report, we will attempt to summarize current information and speculation concerning the involvement of immunoglobulin genes and oncogenes in human B-cell neoplasms, with particular emphasis on work from our own laboratories.

Studies on Burkitt Lymphoma

It has been known for some time that in most cases of Burkitt's lymphoma there is characteristic reciprocal chromosomal transloca- tion between chromosomes 8 and 14 (19,32). In a minority of cases, there is a similar translocation involving chromosomes 8 and 22 or chromosomes 8 and 2 (29,17). Because the immunoglobulin heavy chain genes and light chain genes had been mapped to those areas of chro- mosomes 14, 22 and 2 that were involved in these translocations in the Burkitt tumor (1,9,16,20), it was suggested that these genes might play a significant role in the pathogenesis of this neoplasm (9,16). Subsequently, one of the human homologues of the retroviral oncogenes, the so-called c-*myc* oncogene that is homologous to the avian virus oncogene v-*myc*, was mapped to the terminal portion of the long arm of chromosome 8, the region involved in the typical t(8;14) translocation (18,4,21). This finding suggested that the c-*myc* oncogene might also have an important role in the development of Burkitt's lymphoma.

The immunoglobulin chain locus and the c-*myc* oncogene. A number of laboratories have now studied the structure and function of the immunoglobulin heavy chain locus and the c-*myc* oncogene in a number of Burkitt tumor cell lines with the t(8;14) chromosome transloca- tion. Using a combination of cytogenetic and somatic genetic methods, it has been shown that in these cell lines the immunoglo- bulin heavy chain locus is typically split, with a portion of the locus translocated to the involved chromosome 8, and genes for the constant regions of the heavy chains retained on chromosome 14 (26). Similar methods were used to show that the segment of chromosome 8 translocated to the long arm of chromosome 14 contained the c-*myc* proto-oncogene (18,4,21). In most of the cell lines studied, the c-*myc* gene was found to be arranged in a head-to-head (5' to 5') association. There does, however, appear to be heterogeneity in the breakpoint on both chromosomes 8 and 14 in cell lines derived from

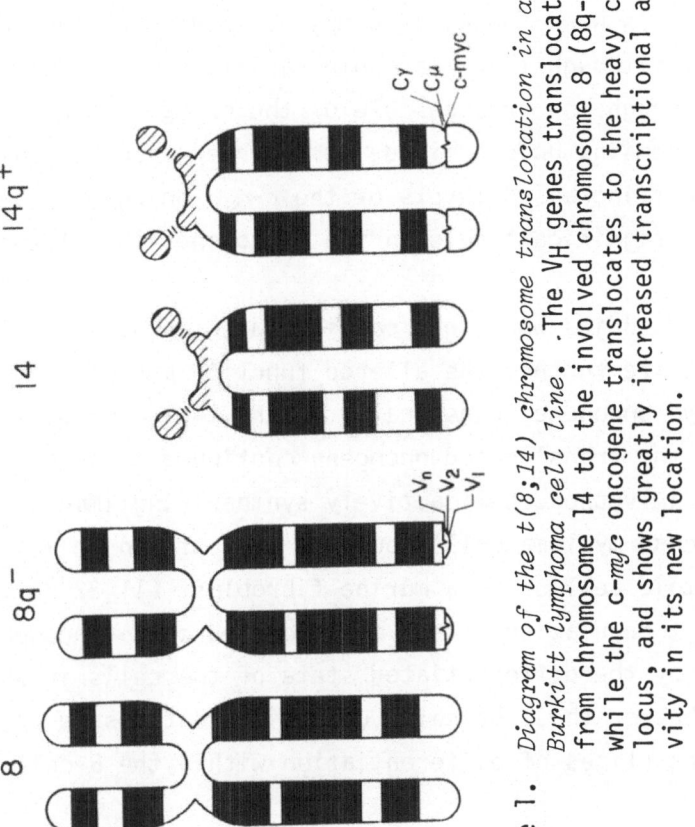

Figure 1. *Diagram of the t(8;14) chromosome translocation in a Burkitt lymphoma cell line.* .The V$_H$ genes translocate from chromosome 14 to the involved chromosome 8 (8q–), while the c-*myc* oncogene translocates to the heavy chain locus, and shows greatly increased transcriptional activity in its new location.

different Burkitt tumors (21,10). Figure 1 is a diagram of the relationship of the immunoglobulin and c-*myc* genes in the P3HR-1
Burkitt lymphoma cell line that contains the t(8;14) translocation.

Subsequent to these findings, additional studies were done to determine the levels of transcription of the c-*myc* oncogene in Burkitt
lymphoma cell lines with this chromosome translocation. In general,
elevated levels were found, as compared to non-neoplastic B-cells
in culture, although there was some variation from case to case
(5,11). Although the precise role of the oncogene product in growth
regulation remains undetermined, it has been assumed from these
findings that increased levels of the c-*myc* oncogene product probably play a significant role in the pathogenesis of these lymphomas.

A number of studies were then carried out with somatic cell hybrids
to characterize further the altered function and altered regulation
of the c-*myc* oncogene translocated to the 14q+ chromosome. It was
found that the translocated oncogene continued to be expressed when
present in a lymphoid cell actively synthesizing immunoglobulin
(i.e., a mouse myeloma cell), but was shut off in an immature human
lymphoblastoid cell or in a murine fibroblast (11,22,23). Thus it
appears that regulation of the translocated c-*myc* oncogene can be
determined by the differentiated state of the cells in which it
resides, whether these be cells of a different tissue or even cells
at different stages of differentiation within the B-cell lineage.

As a result of these various observations, we have speculated that
increased c-*myc* expression in Burkitt's lymphoma may depend on
"enhancer-like" genetic elements that are active in immunoglobulin-
secreting cells such as the Burkitt tumor cells and plasma cells,
but not in less differentiated lymphoblastoid cells. Because the
involved c-*myc* oncogene appears to be activated even if it is
located more than 50 kb from the immunoglobulin heavy chain locus,

we have suggested that these putative differentiation stage-speci-
fic enhancer elements might be termed "long-range enhancers" (23).
Such elements would appear to be different from the enhancer known
to be located between the switch and the J_H regions of the heavy
chain gene, since this enhancer is known to function in lymphoblastoid cells and also seems to act over a shorter range. We are
currently attempting to extend these two concepts, of limited
"windows" of differentiation in which oncogenes may function, and
of "long-range enhancers" that can influence that function, to a
number of other human hematopoietic neoplasms in which different
nonrandom chromosomal translocations have been observed.

The immunoglobulin light chain loci and the c-*myc* oncogene. Several laboratories have now extended the studies on Burkitt lymphoma
cell lines to those cases in which there is a chromosome translocation involving chromosomes 8 and 22 or chromosomes 2 and 8 (3,6,2,
12). As with the more common t(8;14) translocation, the results in
these cell lines indicate a role for both the c-*myc* oncogene and
an immunoglobulin gene locus, but the mechanism whereby the proto-
oncogene is activated appears to be different. Several cell lines
have now been studied that have a translocation between the long
arm of chromosome 8 and the long arm of chromosome 22, using both
somatic cell genetics and *in situ* hybridization techniques (3,6,
7). Our data indicate that in these cells the lambda light chain
locus on chromosome 22 is interrupted, with translocation of the
constant region sequences to chromosome 8, where the c-*myc* gene
remains in an unrearranged configuration. Thus, in this translocation, the immunoglobulin gene sequences are brought to the c-*myc*
gene in a head-to-tail configuration, with the 5' end of the C-
lambda sequences facing the 3' end of the oncogene locus. Figure 2
is a diagram of the t(8;22) translocation in a Burkitt lymphoma
cell line.

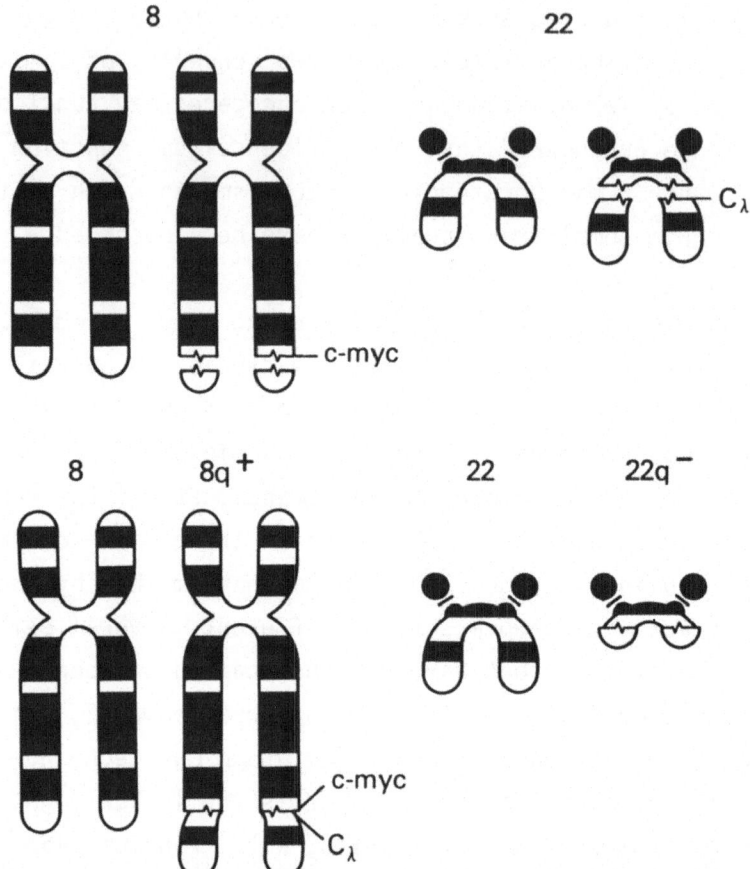

Figure 2.   *Diagram of the t(8;22) translocation observed in BL2*
Burkitt lymphoma cells.   The c-*myc* oncogene remains on
chromosome 8, while the C-lambda gene locus translocates to a region distal to it.

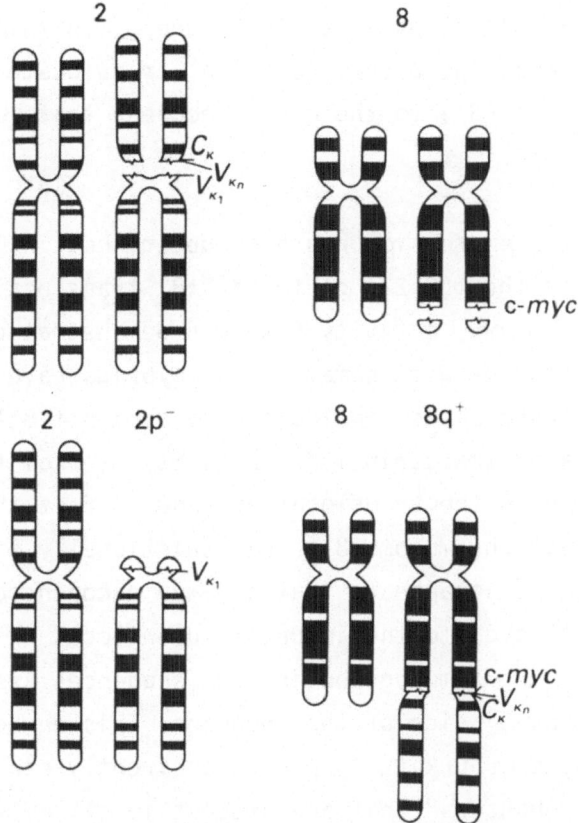

Figure 3. *Diagram of the t(2;8) translocation in JI Burkitt lymphoma cells.* The c-*myc* oncogene remains on chromosome 8,
while the C-kappa gene translocates to a region distal
to it.

Similar results have been observed in studies of Burkitt cells with
the other variant chromosome translocation involving the short arm
of chromosome 2 and the long arm of chromosome 8 (2,12). As in the
t(8;22) translocation, the results indicate that the breakpoint on
chromosome 8 is distal to the c-*myc* oncogene, and that the breakpoint on the short arm of chromosome 2 is within the immunoglobulin
light chain locus. As illustrated in Figure 3, in this case it is
the kappa light chain gene sequences that translocate to a chromosomal region distal (3') to the c-*myc* oncogene that remains on
chromosome 8.

The level of c-*myc* expression has been determined in Burkitt cell
lines with either the t(8;22) or the t(2;8) translocation, and
enhanced transcriptional activity has been demonstrated (6,2).
Interestingly, studies with somatic cell hybrids have indicated
that in all of these cases, including the common t(8;14) translocation, the increased transcriptional activity is from the c-*myc*
oncogene involved in the translocation, and in fact the c-*myc* oncogene on the normal chromosome 8 is transcriptionally silent. Under
these circumstances it appears that a c-*myc* oncogene can be activated whether it moves to the immunoglobulin locus or vice versa,
and also whether the immunoglobulin gene sequences are located on
the 5' side or the 3' side of the oncogene. This deregulation of
the c-*myc* oncogene in Burkitt lymphomas apparently can be affected
by "long-range enhancers" that are present in all of the three
immunoglobulin gene loci.

Other human B-cell lymphomas

The findings with the Burkitt tumor cell lines have suggested that
similar oncogene activating mechanisms might be operative in other
human B-cell neoplasms. It has been recognized for some time that
translocations involving chromosomes 11 and 14 and chromosomes 14
and 18 are common in a variety of human lymphoid tumors, with the

breakpoint on chromosome 14 in the same region as in the Burkitt tumor translocations (8,25,30,15). This suggested possible involvement of the immunoglobulin heavy chain locus, but there has been no candidate oncogene mapped to the relevant regions of the long arms of chromosomes 11 and 18 that might be considered for activation in the same fashion as the c-*myc* oncogene in the Burkitt tumor cells.

We have already begun to study the t(11;14) and t(14;18) translocations in several types of human lymphoid tumors. We have cloned the chromosomal breakpoint from the neoplastic cells of a case of chronic lymphocytic leukemia with the t(11;14) translocation, and have found that the breakpoint on chromosome 14 is indeed within the immunoglobulin heavy chain locus (31). In addition, we have found that rearranged DNA sequences from chromosome 11 have been translocated immediately adjacent to the breakpoint on the involved chromosome 14 (Fig. 4), and this same rearranged DNA segment has been identified in the cells of several other B-cell neoplasms carrying the t(11;14) translocation (13). Because this altered DNA segment was not present in B-cell tumors without the t(11;14) translocation or in non-neoplastic human lymphoblastoid cells, we have suggested that it may represent an as yet unrecognized oncogene, normally located on chromosome 11, that plays a significant role in the pathogenesis of B-cell lymphomas when translocated to the immunoglobulin heavy chain locus on chromosome 14 (31,13).

Similar studies have been initiated with human B-cell neoplasms carrying the t(14;18) translocation. Again, the results indicate that the breakpoint on chromosome 14 interrupts the immunoglobulin heavy chain locus (28). A clone from the joining between chromosome 14 and 18 contains DNA segments specific for chromosome 18, and analysis with this probe indicates that the same segment of chromosome 18 is rearranged in a significant proportion of follicular lymphomas (24). Taken together, these initial results with

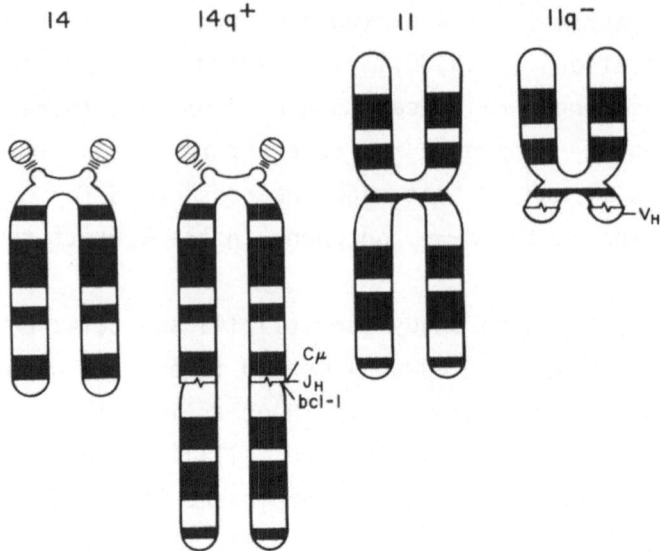

Figure 4.   *Diagram of the t(11;14) translocation occurring in human B-cell neoplasia.*

neoplasms having either the t(11;14) or the (14;18) chromosome translocations suggest that in addition to the c-*myc* oncogene, two other loci, which we have termed bcl-1 and bcl-2, may be important in the pathogenesis of human B-cell neoplasia. Because the same translocations have also been occasionally recognized in T-cell tumors (27), these putative oncogenes may also play a role in these disorders. The current availability of cDNA probes for these two loci, on chromosomes 11 and 18, respectively, should allow investigation of their structure, organization, and expression, and may provide additional clues to the mechanisms of gene activation in neoplastic lymphocytes. Ultimately, such probes may also find practical clinical uses in diagnosis and patient management.

SUMMARY

Recent investigations of cell lines from Burkitt's lymphoma have indicated that the chromosomal translocations observed in this tumor consistently bring the c-*myc* oncogene on chromosome 8 adjacent to a rearranged and transcriptionally active immunoglobulin gene, with resultant "activation" of the oncogene. In other B-cell lymphomas, with translocations involving chromosomes 11 and 14 or chromosomes 14 and 18, a related phenomenon may occur involving as yet unidentified "oncogenes" on chromosomes 11 and 18, and the immunoglobulin heavy chain locus on chromosome 14. Combined techniques involving somatic cell genetics, molecular genetics, and cytogenetics are providing important new knowledge on the biology of neoplasia and opportunities for additional advances in the immediate future.

REFERENCES

1) CROCE, C.M. *et al.* (1979). Chromosomal location of the human immunoglobulin heavy chain genes. Proc. Natl. Acad. Sci. USA, 76, 3416.

2) CROCE, C.M. *et al.* (1983). Transcriptional activation of an unrearranged and untranslocated c-*myc* oncogene by translocation of a C-lambda locus in Burkitt lymphoma cells. Proc. Natl. Acad. Sci. USA, 80, 6922.

3) CROCE, C.M. *et al.* (1984). The translocated c-*myc* gene of Burkitt lymphoma is transcribed in plasma cells and repressed in lymphoblastoid cells. Proc. Natl. Acad. Sci. USA, 81, 3170.

4) DALLA FAVERA, R. *et al.* (1982). Assignment of the human c-*myc* onc-gene to the region of chromosome 8 which is translocated in Burkitt lymphoma cells. Proc. Natl. Acad. Sci. USA, 79, 7824.

5) DALLA FAVERA, R. *et al.* (1983). Translocation and rearrangements of the c-*myc* onc-gene in human differentiated B-cell lymphomas. Science, 219, 93.

6) DE LA CHAPELLE, A. *et al.* (1983). Lambda Ig constant region genes are translocated to chromosome 8 in Burkitt's lymphoma with t(8;22). Nucleic Acids Res., 1, 133.

7) EMANUEL, B. *et al.* (1984). The 2p breakpoint of a 2;8 translocation in Burkitt's lymphoma interrupts the V_k locus. Proc. Natl. Acad. Sci. USA, 81, 2444.

8) EMANUEL, B. *et al.* (1984). Non-identical 22q11 breakpoint for the t(9;22) of CML and the t(8;22) of Burkitt's lymphoma. Cytogenet. Cell Genet., 38, 127.

9) ERIKSON, J., MARTINIS, J. & CROCE, C.M. (1981). Assignment of the human genes for immunoglobulin chains to chromosome 22. Nature, 294, 173.

10) ERIKSON, J. *et al.* (1982). Translocation of immunoglobulin V_H genes in Burkitt lymphoma. Proc. Natl. Acad. Sci. USA, 79, 5611.

11) ERIKSON, J. *et al.* (1983). Transcriptional activation of the c-*myc* oncogene in Burkitt lymphoma. Proc. Natl. Acad. Sci. USA, 80, 820.

12) ERIKSON, J. *et al.* (1983). Translocation of an immunoglobulin kappa locus to a region 3' of an unrearranged c-*myc* oncogene enhances c-*myc* transcription. Proc. Natl. Acad. Sci. USA, 80, 7851.

13) ERIKSON, J. *et al.* (1984). The chromosome 14 breakpoint in neoplastic B cells with the t(11;14) translocation involves the immunoglobulin heavy chain locus. Proc. Natl. Acad. Sci. USA, 81, 4144.

14) FINAN, J. *et al.* (1978). Cytogenetics of chronic T cell leukemia, including two patients with a 14q+ translocation. Virchows Archiv B Cell Pathol., 29, 121.

15) FUKUHARA, S. *et al.* (1979). Chromosome abnormalities in poorly differentiated lymphocytic leukemia. Cancer Res., 39, 3119.

16) KLEIN, G. (1981). The role of gene dosage and genetic trans-positions in carcinogenesis. Nature, 294, 313.

17) LENOIR, G.M. *et al.* (1982). Correlation between immunoglobulin light chain expression and variant translocation in Burkitt's lymphoma. Nature, 298, 474.

18) MALCOLM, S. *et al.* (1982). Localization of human immunoglobulin light chain variable region genes to the short arm of chromosome 2 by *in situ* hybridization. Proc. Natl. Acad. Sci. USA, 79, 4957.

19) MANOLOV, G. & MANOLOVA, Y. (1972). Marker band in one chromosome 14 from Burkitt lymphoma. Nature, 237, 33.

20) McBRIDE, O.W. *et al.* (1982). Chromosomal location of human kappa and lambda immunoglobulin light chain constant region genes. J. Exp. Med., 155, 1480.

21) NEEL, B. *et al.* (1982). Two human c-*onc* genes are located on the long arm of chromosome 8. Proc. Natl. Acad. Sci. USA, 79, 7842.

22) NISHIKURA, K. *et al.* (1983). Differential expression of the normal and of the translocated human c-*myc* oncogenes in B-cells. Proc. Natl. Acad. Sci. USA, 80, 4822.

23) NISHIKURA, K. *et al.* (1984). Repression of recombinant *mu* gene and translocated c-*myc* in mouse 3T3 cell in Burkitt lymphoma cell hybridomas. Science, 224, 399.

24) PEGORARO, L. *et al.* (1984). A t(14;18) and a t(8;14) chromosome translocation in a cell line derived from an acute B-cell leukemia. Proc. Natl. Acad. Sci. USA, 81, 7166.

25) ROWLEY, J.D. & TESTA, J.R. (1982). Chromosome abnormalities in malignant hematologic diseases. In: Adv. Cancer Res., 36, 103.

26) TAUB, R. *et al.* (1982). Translocation of the c-*myc* gene into the immunoglobulin heavy chain. Proc. Natl. Acad. Sci. USA, 79, 7839.

27) TSUJIMOTO, Y. *et al.* (1984). Cloning of the chromosome break-point of neoplastic B-cells with the t(14;18) chromosome translocation. Science, 226, 1097.

28) TSUJIMOTO, Y. *et al.* (1984). Molecular cloning of the chromo-
    somal breakpoint of B-cell lymphomas and leukemias with the
    t(11;14) chromosome translocation. Science, 224, 1403.

29) VAN DEN BERGHE, H. *et al.* (1979). Variant translocation in
    Burkitt lymphoma. Cancer Genet. Cytogenet., 1, 9

30) YUNIS, J.J. (1983). The chromosomal basis of human neoplasia.
    Science, 221, 227.

31) YUNIS, J.J. *et al.* (1982). Distinctive chromosomal abnormal-
    ities in histologic subtypes of non-Hodgkin's lymphoma.
    N. Engl. J. Med., 307, 1231.

32) ZECH, L. *et al.* (1976). Characteristic chromosomal abnormal-
    ities in biopsied and lymphoid cell lines from patients with
    Burkitt and non-Burkitt lymphoma. Int. J. Cancer, 17, 47.

THE FUNCTIONS OF ONCOGENE PRODUCTS

Tony Hunter

Molecular Biology and Virology Laboratory
The Salk Institute, Post Office Box 85800
San Diego, California 92138, USA

An understanding of the actions of oncogenes in the process of malignant transformation will ultimately depend on a knowledge of the functions of the protein products of these genes. To date about twenty different oncogenes of either viral or tumor origin have been identified. One or more protein products of each of these onco-genes have been described. Some progress has been made in assigning functions to oncogene products, and at least one function has been ascribed to over half these proteins. I propose to review briefly our current knowledge in this area.

From the outset it should be realised that a detailed understanding of structure and function for any of these proteins is lacking. All of the oncogenes I will discuss, however, have been molecularly cloned, and from the nucleotide sequences of the coding regions of these clones predicted protein structures have been derived for each oncogenic protein. As will be seen this information has been extremely useful in classification. It should also be appreciated that even where particular functions have been identified, it is entirely possible that these proteins are multifunctional.

Table of Oncogenes

ONCOGENE	TRANSFORMING PROTEIN	LOCATION	FUNCTION
GROUP 1			
v-*src*	pp60v-*src*	Plasma membrane	Tyr PK
v-*yes*	P90*gag-yes*	Plasma membrane?	Tyr PK
v-*fgr*	P70*gag-fgr*	?	Tyr PK
v-*fps*	P140*gag-fps*	Cytoplasmic	Tyr PK
v-*fes*	P85*gag-fes*	Cytoplasmic	Tyr PK
v-*abl*	P160*gag-abl*	Plasma membrane?	Tyr PK
v-*ros*	P68*gag-ros*	?	Tyr PK
v-*erb*-B	gp65/74*erb*B	Cytoplasmic membranes	Tyr PK (EGF.R)
GROUP 1A			
v-*fms*	gP180*gag-fms*	Cytoplasmic membranes	PK?
v-*mil*	P100*gag-mil*	Cytoplasm	PK?
v-*raf*	P90*gag-raf*	Cytoplasm	PK?
v-*mos*	p37*env-mos*	Soluble cytoplasm	PK?
GROUP 2			
v-*sis*	gP28*env-sis*	Secreted	Growth factor(PDGF?)
GROUP 3			
H-*ras*	p21H-*ras*	Plasma membrane	GTP binding/GTPase
K-*ras*	p21K-*ras*	Plasma membrane	GTP binding/GTPase
N-*ras*	p21N-*ras*	Plasma membrane	GTP binding/GTPase
GROUP 4			
v-*myc*	P110*gag-myc*	Nuclear matrix	DNA binding
v-*myb*	P48*gag-myb-env*	Nucleus	?
v-*fos*	p55v-*fos*	Nucleus	?
v-*ski*	P125*gag-ski*	Nucleus	?
v-*ets*	P135*gag-myb-ets*	?	?
v-*erb*-A	P75*gag-erb*A	Cytoplasm	?
v-*rel*	P56*env-rel*	?	?

Abbreviations: Tyr PK = protein-tyrosine kinase; PK = protein kinase; EGF.R = EGF receptor.
The v-*fps* and v-*fes* genes have been shown to be the avian and feline equivalents of the same cellular gene. This is also true for the v-*mil* (avian) and v-*raf* (murine) gene pair and probably the v-*yes* (avian) and v-*fgr* (feline) gene pair as well. There are H-*ras* and K-*ras* genes of both viral and tumor origin. Therefore these genes have not been given the 'v-' prefix.

GROUP 1 ONCOGENES

The functions of the known oncogene products fall into natural
groups. The largest group contains proteins which have protein-
tyrosine kinase activity (for review see refs. 1 and 2). They
correspond to the products of the v-*src*, v-*yes/fgr*, v-*fps/fes*, v-
abl, v-*ros* and v-*erb*-B viral oncogenes. All these proteins manifest
protein-tyrosine kinase activity *in vitro*. Their structures are
characterized by a stretch of about 250 amino acids whose sequence
is homologous not only to a similar stretch in all other members of
the group, but also to sequences in several serine-specific protein
kinases, including the cAMP-dependent protein kinase catalytic sub-
unit, the cGMP-dependent protein kinase, myosin light chain kinase
and the phosphorylase kinase γ-subunit (for review see ref. 3). In
the case of pp60v-*src* and P140$^{gag-fps}$ a fragment corresponding to
this sequence has been isolated by partial proteolysis and shown to
have protein-tyrosine kinase activity (4-6). When the corresponding
region of P120$^{gag-abl}$ is expressed as part of a hybrid protein in
E. coli it too has phosphotransferase activity (7). For these and
other reasons there seems little doubt that in the cells where they
are found these proteins can and do phosphorylate cellular proteins
on tyrosine. Indeed proteins containing increased levels of phosphotyrosine are readily demonstrated in the pertinent cell types (for
review see ref. 8). One would anticipate that protein-tyrosine phosphorylation can alter protein function, as is the case for protein-
serine/threonine phosphorylation. Aberrant protein-tyrosine phosphorylation mediated by a virally encoded enzyme leading to altered
protein function could induce the transformed phenotype.

Through the use of both conditional and non-conditional viral mutants it has been found that there is a good correlation between
the protein-tyrosine kinase activity of these oncogenic proteins
and their ability to induce the transformed phenotype. Nevertheless
there is no formal proof that this activity is necessary or suffi-

cient for their transforming ability. While it seems likely that
protein phosphorylation is important, to date no tyrosine-phosphorylated substrates have been identified which can be demonstrated
to be crucial for transformation (8). There is the real possibility
that these oncogenic proteins have additional functions. For instance
there are recent reports that pp60V-*src* and P68*gag-ros* can phosphorylate certain lipids (9,10), although whether this is an intrinsic
activity is unclear. The virally encoded protein-tyrosine kinases
range in size from 160 kDa to 68 kDa. Although in most cases some
of this mass is contributed by sequences derived from viral structural genes, each protein kinase has from 30 to 100 kDa of protein
sequence ancillary to its catalytic domain. These additional sequences could act to modulate the phosphotransferase activity either
in a regulatory fashion or else by dictating a specific subcellular
location. Alternatively such regions might have a separate function.

Each of the viral oncogenes has been derived from a homologous cellular gene. Although like most of the oncogenes the v-*onc* genes in
Group 1 only represent part of the corresponding cellular gene
(with the exception of the v-*src* gene), the protein kinase domain
is highly conserved between the respective v-*onc*/c-*onc* gene pairs.
One would anticipate therefore that the products of these cellular
genes would also be protein-tyrosine kinases and indeed this is so
for the products of the c-*src* and c-*fps/fes* genes (11,12). In the
other cases the issue is unresolved. The products of the c-*yes/fgr*
and c-*ros* genes have not yet been identified, while the putative
product of the c-*abl* gene apparently lacks protein kinase activity
(13). In all likelihood the c-*erb*-B gene is the epidermal growth
factor (EGF) receptor gene (14,15). The EGF receptor is a representative of another class of protein-tyrosine kinases whose activities are induced by binding polypeptide ligands (for review see
refs. 2 and 3). So far all the members of this class are peptide
growth factor receptors, including the EGF, the platelet-derived

growth factor (PDGF), the insulin and insulin-like growth factor 1
(IGF-1) receptors. The inducible properties of this type of protein-
tyrosine kinase suggests that protein-tyrosine phosphorylation may
be involved in the response of cells to growth factors. If a growth
control pathway utilising protein-tyrosine phosphorylation exists,
this might provide a point at which the viral protein-tyrosine kinases could intercede to drive cells to grow continuously.

Assuming that protein phosphorylation is important for transformation by Group 1 oncogenes, the protein substrates for these protein-tyrosine kinases are of obvious interest. A number of proteins
containing increased levels of phosphotyrosine have been identified
in cells transformed by the relevant viruses (8), and in receptor-
bearing cells treated acutely with EGF or PDGF (16,17). These proteins are putative primary substrates for the protein-tyrosine
kinases in question. In general there is considerable overlap in
the spectra of substrates in the different types of virally transformed cell, suggesting that these enzymes have rather similar substrate specificities (8). The major substrate proteins are vinculin
(18) (a cytoskeletal protein found in adhesion plaques), p81 (K.
Gould, T. Hunter, J. Cooper & A. Bretscher, unpublished results)
(a protein located in microvillar cores of both fibroblasts and gut
epithelial cells), p50 (19) (a protein which associates with newly
synthesized molecules of pp60^{v-src}), three glycolytic enzymes (20)
(enolase, lactate dehydrogenase, phosphoglycerate mutase) and p36
(21,22) (a protein located in the subplasma membrane cortical
skeleton in fibroblasts). Minor substrates include p42. p42 is the
major substrate for the growth factor receptor protein-tyrosine
kinases (17).

Some generalisations can be made about these proteins. Their unphosphorylated counterparts are relatively abundant proteins (0.05-
0.3% of total cell protein). At steady state the fraction of mole-

cules phosphorylated on tyrosine tends to be low, ranging from 1-15%. The exception to both these rules is p42, which comprises only 0.002% of total cell protein and is phosphorylated to at least 50% in growth factor treated cells.

What are the functions of these substrates and how does tyrosine phosphorylation affect their function? Vinculin has been proposed to act as a linker between the plasma membrane and actin bundles, so that its phosphorylation might lead to the observed disruption of actin bundles in transformed cells. The stoichiometry of phosphorylation, however, is only 1%. This combined with data from cells infected with mutant viruses makes it unlikely that vinculin phosphorylation is the sole cause of the altered morphology of such virally transformed cells. It seems improbable that the tyrosine phosphorylation of the three glycolytic enzymes is of functional consequence for glycolysis since not only is their stoichiometry of phosphorylation less than 10% but none of these enzymes is rate limiting. Possibly these proteins are phosphorylated gratuitously simply because they are rather abundant. p81 and p36 are both plasma membrane-associated and are thus in a position to be phosphorylated by the viral protein-tyrosine kinases, many of which are also found in this cellular compartment. We believe that both proteins have structural functions. Their phosphorylation might play a role in the multiple membrane changes observed in transformed cells, but there is no evidence that these phosphorylations are not incidental.

Phosphotyrosine is a rare modification in normal cells accounting for 0.05% of total phosphate linked to protein (19). This rises to 0.5% in cells transformed by the appropriate viruses (19). The identified major substrates do not account for the total increment in phosphotyrosine in protein in transformed cells. We suspect that several more minor substrates contribute to the increase, and that such proteins are likely to have high stoichiometries of phospho-

rylation and be important for transformation. In many ways p42 has
the hallmarks of a genuine substrate. Little is known about this
protein although it is found primarily in a soluble state in the
cytoplasm. Its phosphorylation is induced by a wide variety of
mitogens (17,23-25). Despite the good correlation between p42 phosphorylation and mitogenesis, however, there is no direct evidence
that its phosphorylation is a prerequisite for progression
through the cell cycle. Furthermore p42 is not detectably phosphorylated in all the expected virally transformed cells, so it is not
clear whether its phosphorylation could be instrumental in virally-
driven growth.

GROUP 1A ONCOGENES

The second group of oncogenes, which includes the v-*fms*, v-*mil*/*raf*
and v-*mos* genes, encode predicted proteins that also have a 250
amino acid stretch with homology to the catalytic domain of the
proteins in Group 1 (for review see ref. 26). The assays used to
demonstrate the protein-tyrosine kinase activity of the Group 1 oncogene products, however, give somewhat equivocal results for the
products of Group 1A oncogenes. The v-*fms* protein is weakly phosphorylated on tyrosine in an *in vitro* reaction (27), while the
v-*mil*/*raf* (28) and the mutant P85$^{gag-mos}$ (29) proteins are phosphorylated only on serine and/or threonine. Cells transformed by
the relevant viruses do not display increases in the level of phosphotyrosine in protein (27,30,31). The structure of the v-*fms* protein is reminiscent of that of the v-*erb*-B protein and it has been
surmised that the v-*fms* gene represents a fragment of another
growth factor receptor gene. It is noteworthy that it has been difficult to demonstrate convincing *in vitro* or *in vivo* protein-tyrosine kinase activity for the v-*erb*-B protein despite its apparent
identity with the catalytic part of the EGF receptor. In this regard the v-*fms* product is similar to the v-*erb*-B product and it
remains possible that the v-*fms* protein is a protein-tyrosine kinase

with much more restricted activity than those in Group 1.

Given the fact that the v-*mil/raf* and v-*mos* proteins have almost as good homology with the protein-serine/threonine kinases as with the protein-tyrosine kinases we cannot dismiss the possibility that these proteins are in fact protein-serine/threonine kinases. *A priori* there is no reason why aberrant protein-tyrosine phosphorylation should be any more effective at inducing the transformed phenotype than protein-serine or threonine phosphorylation since both types of protein modification can in principle alter protein function. One protein-serine kinase which is intimately associated with cell growth is the Ca^{2+}/phospholipid-dependent diacylglycerol activated protein kinase, protein kinase C (for review see ref. 32). It is an outside possibility that the *raf/mil* gene is the protein kinase C gene.

GROUP 2 ONCOGENES

The third group of oncogenes contains only one member, v-*sis*. The v-*sis* gene product corresponds to one of the two chains of PDGF (33,34). The v-*sis* protein is secreted from transformed cells, and acts very like PDGF in its ability to stimulate resting fibroblasts to grow and to stimulate the PDGF receptor protein-tyrosine kinase (35,36). The v-*sis* protein may therefore act via an autocrine system to transform cells by a mechanism involving protein-tyrosine phosphorylation. It is pertinent that only cells expressing PDGF receptors are susceptible to transformation by the v-*sis* gene. The autocrine hypothesis may be an oversimplification, however, because authentic PDGF, unlike the transforming growth factors, by itself is not able to induce the transformed phenotype. The availability of a number of cloned growth factor genes (e.g. EGF, TGF and IGF-1) means that it will be possible to test whether other growth factor genes expressed in a suitable fashion are oncogenic.

GROUP 3 ONCOGENES

The fourth group of oncogenes consists entirely of *ras* genes, which are of both viral and tumor origin. The *ras* oncogenes are derived from three types of related but distinct c-*ras* genes, c-Ha-*ras*, c-Ki-*ras* and c-N-*ras* (for review see ref. 37). The product of all these *ras* genes are 21 kDa proteins. The oncogenic p21^{ras}'s differ from their normal cellular counterparts by one or more single amino acid substitutions (37). All types of p21^{ras} are largely associated with the inner face of the plasma membrane and contain a palmitate moiety covalently attached near the C-terminus, which is probably important for this membrane affiliation. All p21^{ras}'s bind guanine nucleotides very tightly and the affinity of oncogenic p21^{ras}'s is similar to that of normal p21^{ras} (38,39). Purified p21^{v-ras}'s autophosphorylate in the presence of GTP, but have never been observed to phosphorylate exogenous substrates (38). Since neither normal p21^{ras} nor tumor-derived oncogenic p21^{ras} autophosphorylate, this property is probably not important for transformation. Recently it has been found that purified p21^{ras} has GTPase activity (40). Intriguingly the oncogenic p21^{ras}'s have considerably lower GTPase activity than their normal cellular counterparts.

In many ways the properties of p21^{ras} are similar to those of the coupling factors, proteins which are involved in the activation of adenylate cyclase in response to β-adrenergic hormones (for review see ref. 41). Like p21^{ras}, the coupling factors are GTP-binding proteins with GTPase activity which are associated with the inner face of the plasma membrane. The coupling factors bind GTP effectively only in the presence of occupied hormone receptor and can only deliver the stimulatory signal to adenylate cyclase in the GTP-bound state. The slow hydrolysis of the bound GTP to GDP temporarily inactivates the coupling factor until the bound GDP has been exchanged for GTP. It is tempting to speculate that p21^{ras}'s have similar signal-transducing functions, but link an unknown surface

receptor to an unidentified cytoplasmic signal generating system.
If this were true, then, because of their defective GTPase, the
oncogenic p21^{ras}'s would remain active for a proportionately longer
time upon binding GTP than the normal p21^{ras}'s, and provide a continuous rather than a regulated signal. Given the properties of
cells transformed with oncogenic ras genes, a growth factor receptor seems the most likely candidate to be coupled via p21^{ras}. Indeed
there is a preliminary report of a possible interaction between the
EGF receptor and p21^{v-ras} (42).

GROUP 4 ONCOGENES

The fifth group of oncogenes, which includes the v-myc, v-myb, v-
fos and v-ski genes, are grouped on the basis of a common nuclear
localization, rather than on a common function. Their presence in
the nucleus, however, suggests that they might all be able to
affect gene expression directly. The v-myc protein is a DNA binding
protein but displays little sequence specificity (43). A major population of the v-myc protein is tightly associated with the nuclear
matrix (44). The c-myc protein is also found the in nucleus (44).
Its expression is rapidly but transiently induced following treatment of resting cells with mitogens (45,46). Both v-myc and c-myc
proteins have short half lives on the order of 30 min (44). This
property together with the regulated expression of the c-myc gene
in normal cells suggests that the c-myc protein may play an important role in early events in the cell cycle, possibly being involved
in the induction of other genes required for the progression through
the cell cycle. In this regard the numerous examples of translocated
c-myc genes in non-viral cancers are of interest. Many of these
translocated genes are expressed at higher levels than their normal
counterparts. More importantly, however, the transcription of these
genes is not regulated in the proper fashion, so that they are expressed continuously rather than only early in the G1 phase of the

cell cycle. This loss of temporal regulation may be crucial in the
oncogenic potential of both v-*myc* and altered c-*myc* genes.

The predicted v-*myc* protein has a weak sequence homology with the
predicted product of the v-*myb* gene and also with that of the adenovirus E1A gene (47). The latter protein, like the *myc* proteins, is
a short-lived nuclear matrix associated protein. In addition the
E1A protein regulates the transcription of both adenovirus and cellular genes. If the sequence homology between the E1A and *myc* proteins reflects an analogous function, this would be a further indication that *myc* proteins might serve as transcriptional regulators.
The precise subnuclear location of the v-*myb* protein is not known,
but the fact that the v-*myb* protein adopts an extranuclear location
when AMV-infected myeloblasts are induced to differentiate with TPA
(48) suggests that it might not be tightly anchored in the nucleus.

There is no apparent amino acid sequence homology between the products of *fos* genes and either the v-*myc* or v-*myb* proteins. Both the
v-*fos* and c-*fos* proteins are highly modified short-lived nuclear
phosphoproteins (49). Their precise location in the nucleus is not
known, but since they are extracted from nuclei by mild detergent,
they do not appear to be associated with the nuclear matrix. It has
recently been found that treatment of quiescent fibroblasts with a
variety of mitogens leads to a rapid but transient burst of transcription of the c-*fos* gene (50,51). This occurs even in the presence of inhibitors of protein synthesis and appears to precede the
induction of c-*myc* gene transcription. From the foregoing arguments
about *myc* protein function, it appears likely that *fos* proteins
also induce the transcription of other cellular genes in early G1.
Transformation by *fos* genes could again be due to improper temporal
regulation of *fos* gene transcription. The last protein in this
group is the v-*ski* gene, and apart from its nuclear location very
little is known about this protein or its function.

Finally there is a set of oncogenes whose products cannot yet be classified due to insufficient knowledge about their protein products or else because the sequence of the predicted protein does not have obvious homology with any other sequenced protein. Oncogenes in this category are the v-*ets*, v-*erb*-A, v-*rel* and v-*kit* viral oncogenes and a number of recently cloned tumor oncogenes, including the *neu* and *met* genes.

CONCLUSIONS

The limited number of functional themes apparent among the products of the oncogenes identified so far suggests that the number of possible mechanisms of transformation is restricted. For instance protein phosphorylation, although not a universal feature, seems likely to be involved in transformation by over half the known oncogenes. A description of the precise mechanisms will require not only a fuller knowledge of both the oncogenic proteins themselves and their targets but also a deeper understanding of the cellular processes which control growth, shape, movement and metabolism.

REFERENCES

1) BISHOP, J.M. (1983). Cellular oncogenes and retroviruses. Ann. Rev. Biochem., 52, 301.

2) SEFTON, B.M. & HUNTER, T. (1984). Tyrosine protein kinases. Adv. Cycl. Nucl. and Protein Phosphorylation Res., 18, 195.

3) HUNTER, T. & COOPER, J.A. (1985). Protein-tyrosine kinases. Ann. Rev. Biochem., in press.

4) LEVINSON, A.D., COURTNEIDGE, S.A. & BISHOP, J.M. (1981). Structural and functional domains of the Rous sarcoma virus transforming protein (pp60^{src}). Proc. Natl. Acad. Sci. USA, 78, 1624.

5) BRUGGE, J.S. & DARROW, D (1984). Analysis of the catalytic domain of phosphotransferase activity of two avian sarcoma virus transforming proteins. J. Biol. Chem., 259, 4550.

6) WEINMASTER, G., HINZE, E. & PAWSON, T. (1983). Mapping of
    multiple phosphorylation sites within the structural and cata-
    lytic domains of the Fujinami sarcoma virus transforming pro-
    tein. J. Virol.,46, 29.

7) WANG, J.Y.J., QUEEN, C. & BALTIMORE, D. (1982). Expression of
    an Abelson murine leukemia virus-encoded protein in *Escheri-
    chia coli* causes extensive phosphorylation of tyrosine resi-
    dues. J. Biol. Chem., 257, 13181.

8) COOPER, J.A. & HUNTER, T. (1983). Regulation of cell growth
    and transformation by tyrosine-specific protein kinases: The
    search for important cellular substrate proteins. Current
    Topics in Microbiol. and Immunol., 107, 125.

9) SUGIMOTO, Y., WHITMAN, M., CANTLEY, L.C. & ERIKSON, R.L.(1984).
    Evidence that the Rous sarcoma virus transforming gene product
    phosphorylates phosphatidylinositol and diacylglycerol. Proc.
    Natl. Acad. Sci. USA, 81, 2117.

10) MACARA, I.G., MARINETTI, G.V. & BALDUZZI, P.C. (1984). Trans-
    forming protein of avian sarcoma virus UR2 is associated with
    phosphatidylinositol kinase activity: possible role in tumori-
    genesis. Proc. Natl. Acad. Sci. USA, 81, 2728.

11) OPPERMANN, H., LEVINSON, A.D., LEVINTOW, L., VARMUS, H.E. &
    BISHOP, J.M. (1979). Uninfected vertebrate cells contain a
    protein that is closely related to the product of the avian
    sarcoma virus transforming gene (*src*). Proc. Natl. Acad. Sci.
    USA, 76, 1804.

12) MATHEY-PREVOT, B., HANAFUSA, H. & KAWAI, S. (1982). A cellular
    protein is immunologically cross-reactive with and functional-
    ly homologous to the Fujinami sarcoma virus transforming pro-
    tein. Cell, 28, 897.

13) PONTICELLI, A.S., WHITLOCK, C.A., ROSENBERG, N. & WITTE, O.N.
    (1982). *In vivo* tyrosine phosphorylations of the Abelson virus
    transforming protein are absent in its normal cellular homolog.
    Cell, 29, 953.

14) DOWNWARD, J., YARDEN, Y., MAYES, E., SCRACE, G., TOTTY, N., STOCKWELL, P., ULLRICH, A., SCHLESSINGER, J. & WATERFIELD, M. D. (1984). Close similarity of epidermal growth factor receptor and v-*erb*-B oncogene protein sequences. Nature, 307, 521.

15) ULLRICH, A., COUSSENS, L., HAYFLICK, J.S., DULL, T.J., GRAY, A., TAM, A.W., LEE, J., YARDEN, Y., LIBERMAN, T.A., SCHLESSINGER, J., DOWNWARD, J., MAYES, E.L.V., WATERFIELD, M. D., WHITTLE, M. & SEEBURG, P.H. (1984). Human epidermal growth factor receptor cDNA sequence and aberrant expression of the amplified gene in A431 epidermoid carcinoma cells. Nature, 309, 418.

16) HUNTER, T. & COOPER, J.A. (1981). Epidermal growth factor induces rapid tyrosine phosphorylation of proteins in A431 human tumor cells. Cell, 24, 741.

17) COOPER, J.A., BOWEN-POPE, D., RAINES, E., ROSS, R. & HUNTER, T. (1982). Similar effects of platelet-derived growth factor and eipdermal growth factor on the phosphorylation of tyrosine in cellular proteins. Cell, 31, 263.

18) SEFTON, B.M., HUNTER, T., BALL, E.H. & SINGER, S.J. (1981). Vinculin: a cytoskeletal substrate of the transforming protein of Rous sarcoma virus. Cell, 24, 165.

19) HUNTER, T. & SEFTON, B.M. (1980). The transforming gene product of Rous sarcoma virus phosphorylates tyrosine. Proc. Natl. Acad. Sci. USA, 77, 1311.

20) COOPER, J.A., REISS, N.A., SCHWARTZ, R.J. & HUNTER, T. (1983). Three glycolytic enzymes are phosphorylated at tyrosine in cells transformed by Rous sarcoma virus. Nature, 302, 218.

21) ERIKSON, E. & ERIKSON, R.L. (1980). Identification of a cellular protein substrate phosphorylated by the avian sarcoma virus transforming gene product. Cell, 21, 829.

22) RADKE, K., GILMORE, T. & MARTIN, G.S. (1980). Transformation by Rous sarcoma virus: a cellular substrate for transformation -specific protein phosphorylation contains phosphotyrosine. Cell, 21, 821

23) GILMORE, T. & MARTIN, G.S. (1983). Phorbol ester and diacyl-
    glycerol induce protein phosphorylation at tyrosine. <u>Nature</u>,
    <u>306</u>, 487.

24) NAKAMURA, K.E., MARTINEZ, R. & WEBER, M.J. (1983). Tyrosine
    phosphorylation of specific proteins following mitogen stimu-
    lation of chicken embryo fibroblasts. <u>Mol. Cell. Biol.</u>, <u>3</u>,
    380.

25) COOPER, J.A., SEFTON, B.M. & HUNTER, T. (1984). Diverse mito-
    genic agents induce the phosphorylation of two related 42,000
    dalton proteins on tyrosine in quiescent chick cells. <u>Mol.
    Cell. Biol.</u>, <u>4</u>, 30.

26) BISHOP, J.M. & VARMUS, H.E. (1984). In: "RNA Tumor Viruses"
    (second edition and supplement), R. Weiss, N. Teich, H. Varmus
    & J. Coffin, eds., Chapter 9.

27) BARBACID, M. & LAUVER, A.V. (1981). Gene products of McDonough
    feline sarcoma virus have an *in vitro*-associated protein kinase
    that phosphorylates tyrosine residues: lack of detection of
    this enzymatic activity *in vivo*. <u>J. Virol.</u>, <u>40</u>, 812.

28) MOELLING, K., HEIMANN, B., BEIMLING, P., RAPP, U.R. & SANDER,
    T. (1984). Purified *gag-mil* and *gag-raf* proteins phosphorylate
    serine and threonine in contrast to the tyrosine-specific pro-
    tein kinase *gag-fps*. <u>Nature</u>, in press.

29) KLOETZER, W.S., MAXWELL, S.A. & ARLINGHAUS, R.B. (1983).
    P85*gag-mos* encoded by ts110 Moloney murine sarcoma virus has
    an associated protein kinase activity. <u>Proc. Natl. Acad. Sci.
    USA</u>, <u>80</u>, 412.

30) SEFTON, B.M., HUNTER, T., BEEMON, K. & ECKHART, W. (1980).
    Phosphorylation of tyrosine is essential for cellular trans-
    formation by Rous sarcoma virus. <u>Cell</u>, <u>20</u>, 807.

31) RAPP, U.R., REYNOLDS, F.H. & STEPHENSON, J.R. (1983). New
    mammalian transforming retrovirus: demonstration of polypro-
    tein gene product. <u>J. Virol.</u>, <u>45</u>, 914.

32) NISHIZUKA, Y. (1984). The role of protein kinase C in cell surface signal transduction and tumour production. Nature, 308, 693.

33) DOOLITTLE, R.F., HUNKAPILLER, M.W., HOOD, L.E., DEVARE, S.G., ROBBINS, K.C., AARONSON, S.A. & ANTONIADES, H.N. (1983). Simian sarcoma virus *onc* gene, v-*sis*, is derived from the gene (or genes) encoding a platelet-derived growth factor. Science, 221, 275.

34) WATERFIELD, M.D., SCRACE, G.T., WHITTLE, N., STROOBANT, P., JOHNSSON, A., WASTESON, A., WESTERMARK, B., HELDIN, C.-H., HUANG, J.S. & DEUEL, T.F. (1983). Platelet-derived growth factor is structurally related to the putative transforming protein p28^{sis} of simian sarcoma virus. Nature, 304, 35.

35) DEUEL, T.F., HUANG, J.S., HUANG, S.S., STROOBANT, P. & WATERFIELD, M.D. (1983). Expression of a platelet-derived growth factor-like protein in simian sarcoma virus trans-formed cells. Science, 221, 1348.

36) ROBBINS, K.C., ANTONIADES, H.N., DEVARE, S.G., HUNKAPILLER, M.W. & AARONSON, S.A. (1983). Structural and immunological similarities between simian sarcoma virus gene product(s) and human platelet-derived growth factor. Nature, 305, 605.

37) LAND, H., PARADA, L.F. & WEINBERG, R.A. (1983). Cellular onco-genes and multistep carcinogenesis. Science, 222, 771.

38) SHIH, T.Y., PAPAGEORGE, A.G., STOKES, P.E., WEEKS, M.O. & SCOLNICK, E.M. (1980). Guanine nucleotide-binding and auto-phosphorylating activities associated with the P21^{src} protein of Harvey murine sarcoma virus. Nature, 287, 686.

39) FINKEL, T., DER, C.J. & COOPER, G.M. (1984). Activation of *ras* genes in human tumors does not affect guanine nucleotide binding properties of p21. Cell, 37, 151.

40) McGRATH, J.P., CAPON, D.J., GOEDDEL, D.V. & LEVINSON, A.D. (1984). Comparative biochemical properties of normal and activated human *ras* p21 protein. Nature, 310, 644.

41) GILMAN, A.G. (1984). G proteins and dual control of adenylate cyclase. Cell, 36, 577.

42) KAMATA, T. & FERAMISCO, J.R. (1984). Epidermal growth factor stimulates guanine nucleotide binding activity and phosphorylation of *ras* oncogene products. Nature, 310, 147.

43) MOELLING, K., BUNTE, T., GREISER-WILKE, I., DONNER, P. & PFAFF, E. (1984). Properties of the avian viral transforming proteins *gag-myc*, *myc* and *gag-mil*. Cancer Cells Vol. 2, Oncogenes and Viral Genes, p. 173, Cold Spring Harbor.

44) EISENMAN, R.N., TACHIBANA, C.Y., ABRAMS, H.D. & HANN, S.R. (1984). v-*myc* and c-*myc* encoded protein are associated with the nuclear matrix. Mol. Cell. Biol., in press.

45) KELLY, K., COCHRAN, B.H., STILES, C.D. & LEDER, P. (1983). Cell-specific regulation of the c-*myc* gene by lymphocyte mitogens and platelet-derived growth factor. Cell, 35, 603.

46) CAMPISI, J., GRAY, H.E., PARDEE, A.B., DEAN, M. & SONENSHEIN, G.E. (1984). Cell-cycle control of c-*myc* but not c-*ras* expression is lost following chemical transformation. Cell, 36, 241.

47) RALSTON, R. & BISHOP, J.E. (1983). The protein products of the *myc* and *myb* oncogenes and adenovirus E1A are structurally related. Nature, 306, 803.

48) KLEMPNAUER, K.-H., SYMONDS, G., EVAN, G.I. & BISHOP, J.M. (1984). Subcellular localization of proteins encoded by oncogenes of avian myeloblastosis virus and avian leukemia virus E26 and by the chicken c-*myb* gene. Cell, 37, 537.

49) CURRAN, T., MILLER, A.D., ZOKAS, L. & VERMA, I.M. (1984). Viral and cellular *fos* proteins: a comparative analysis. Cell, 36, 259.

50) GREENBERG, M.E. & ZIFF, E.B. (1984). Stimulation of 3T3 cells induces transcription of the c-*fos* proto-oncogene. Nature, 311, 433.

51) KRUIJER, W., COOPER, J.A., HUNTER, T., & VERMA, I.M. (1984). PDGF induces rapid but transient expression of the c-*fos* gene. Nature, in press.

IDENTIFICATION AND LOCALIZATION OF PHOSPHOPROTEINS IN v-onc TRANSFORMED FIBROBLASTS BY MEANS OF PHOSPHOTYROSINE ANTIBODIES

P.M. Comoglio, D. Cirillo, M.F. Di Renzo
R. Ferracini, F.G. Giancotti, S. Giordano
L. Naldini, G. Tarone & P.C. Marchisio

Institute of Histology, University of Torino
Medical School, C.so M. D'Azeglio 52
10126 Torino, Italy

INTRODUCTION

The transformation process induced by several retroviruses, including Rous sarcoma virus (RSV), Feline sarcoma virus (FeSV), Fujinami avian sarcoma virus (FSV) and Abelson murine leukemia virus (AMuLV), is triggered and maintained by the action of v-*onc* genes which all code for transforming proteins with associated tyrosine kinase activity (for review see 1). Since protein phosphorylation seems to be invariably associated with molecular mechanism(s) involved in growth control and in the neoplastic transformation triggered by these retroviruses, the identification of cellular proteins phosphorylated at tyrosine residues is an issue of major importance. Putative substrates of tyrosine kinases have been identified by means of conventional techniques such as bidimensional separation of total cellular proteins followed by phosphoaminoacid analysis. However, these techniques seem to have intrinsic limitations - as shown also by the failure to identify well known substrates - such as the transforming proteins themselves, which are known to be heavily tyrosine-phosphorylated. The difficulties are generated by

the fact that phosphotyrosine represents less than 2% of phosphoaminoacids also in transformed cells (being less than 0.2% in normal cells); in addition it has been shown that, only a minor fraction, i.e. less than 10%, of each substrate molecules of v-onc
coded kinases, is phosphorylated at tyrosine even in fully transformed cells.

Antibodies able to recognize phosphotyrosine have been employed
in a variety of immunochemical procedures (such as immunodecoration of electroblotted proteins, immunoprecipitation and immunolocalization at both light and electron microscope level) in order
to separate and to localize tentatively - among phosphorylated
cellular proteins - only those quantitatively phosphorylated at
tyrosine residues. A rather broad spectrum of mammalian cells
either normal or transformed by different v-*onc* coded tyrosine-
kinases has been taken into account in this investigation.

A considerable deal of our experiments has been devoted to study
phosphotyrosine-containing proteins in their association with the
cytoskeleton of transformed cells. This approach was adopted for
several reasons among which the fact that most tyrosine kinases
coded by v-*onc* remain associated and are active within the detergent insoluble fraction of cells (e.g. ref. 10) thus suggesting
that the cytoskeleton is a possible target. Another reason was
that previous studies and particularly those done on RSV transformed
cells had indicated that a major location of the pp60 coded by v-
src is the site of membrane-cytoskeleton interaction and particularly at surface-substratum adhesion areas (31).

PRODUCTION OF ANTIBODIES CROSS-REACTING WITH PHOSPHOTYROSINE AND SPECIFICITY TESTS

Phosphotyrosine (P-TYR) antibodies were obtained by immunizing
rabbits against a cross-reacting synthetic hapten (azobenzylphos-

phonate, ABP), linked to keyhole limpet haemocyanin (2-3). The anti-
ABP activity in immune sera and the specificity of purified P-TYR
antibodies were tested in a solid-state binding radioimmunoassay
(5), by measuring the extent of the inhibition exerted by a variety
of low or high molecular weight phosphate containing compounds. The
binding to ABP-BSA-coated polyvinyl wells was completely inhibited
by the hapten; a significant inhibition was also obtained with
phenylphosphate, sterically very similar to the phosphorylated form
of tyrosine. Phosphotyrosine itself, as expected, inhibited P-TYR
antibody binding to a similar extent. On the contrary, the two naturally occurring phosphoaminoacids, phosphoserine and phosphothreo-
nine, did not react significantly with the antibodies. A variety of
other phosphate containing inorganic and organic compounds, including phosphoproteins such as phosvitin, phosphocasein or DNA and RNA,
were also ineffective in inhibiting ABP-antibody binding (7).

The extent of inhibition exerted by phosphotyrosine occurring in
intact cells was evaluated in a similar assay, by measuring the
inhibition of the hapten- antibody binding operated by cell extracts solubilized by SDS-DOC-Triton and extracted by phenol. Protein extracts prepared from cells transformed by v-*onc* coded tyrosine kinases inhibited three to five fold more than similar preparations obtained from control untransformed cells (4).

For further studies, antibodies were purified by affinity chromatography on ABP-BSA coupled to cyanogen bromide activated Sepharose
4B. Bound antibodies were eluted with phenyl phosphate and dialyzed
against PBS, as described (7).

IMMUNODECORATION OF ELECTROBLOTTED PROTEINS

Mouse and rat fibroblasts, either control or transformed by different retroviruses, were extracted by the non-ionic detergent NP40
(1%), as previously described (7), in order to prepare detergent

insoluble cell fractions known to contain most of the v-*onc* coded
tyrosine kinases (e.g. 10). The detergent-insoluble fraction -
which included nuclear chromatin components, cytoskeletal proteins
and intra- and extracellular molecules associated with the plasma
membrane - was further solubilized in SDS-DOC, run on a SDS-PAGE
gradient slab gel and electroblotted onto pure nitrocellulose membrane according to Towbin (8) and Burnette (9). The membrane was
then decorated with P-TYR antibodies.

A series of non-ionic detergent insoluble proteins were decorated
by P-TYR antibodies. It was observed that the Mr of proteins solubilized from cells transformed by kinases coded by different v-*onc*
genes differed significantly (4).

P-TYR antibodies decorated mainly the autophosphorylated kinases,
as identified by Mr and appropriate antisera when available. Additional major bands were observed in v-*src* and in v-*abl* transformed
fibroblasts. Besides the kinases, two proteins of approx. Mr 130
and 70 kd were revealed in both v-*src* transformed rat and mouse
fibroblasts (7). In v-*abl* transformed fibroblasts a major protein
of 70 kd was identified in addition to the 120 kd kinases. In cells
transformed by other v-*onc* (v-*fes* and v-*fps*) the kinases were
accompanied only by minor bands of 70 and 55 kd, decorated by P-
TYR antibodies (4).

IMMUNOPRECIPITATION OF DETERGENT INSOLUBLE CELL PROTEINS

In order to demonstrate that electroblotted proteins decorated by
P-TYR antibodies were actually phosphorylated at tyrosine, detergent insoluble proteins were phosphorylated *in vitro* by ^{32}P- γ-ATP
- according to Burr (10) and Gacon (11) - and immunoprecipitated.
The phosphorylation *in situ* was possible since it had already been
shown that several v-*onc* coded kinases are associated with the cell

fraction insoluble in non-ionic detergents. The detergent-insoluble
fraction was then dissolved in Laemmli buffer and separated in SDS-
PAGE or solubilized for immunoprecipitation in RIPA buffer.

In this set of experiments it was found that tyrosine kinases coded
by different oncogenes could actually phosphorylate the detergent
insoluble fraction at different extent. This result agrees with
previously reported data about the association of the different
oncogene coded kinases with the cytoskeletal fraction (10,12-13).
For instance, detergent insoluble proteins were strongly phosphorylated by pp60*src*; on the contrary a lower degree of phosphorylation
was observed in cells transformed by the v-*fps* coded p130, in accordance with its known partial (50%) detergent solubility (13). No
P-TYR proteins were precipitated from parental non-transformed cell
lines. When rat and mouse cells transformed by v-*src* were compared,
P-TYR proteins of similar Mr were found (7). On the contrary, the
patterns of proteins precipitated by P-TYR antibodies from the same
cell lines, transformed by different retroviruses, differed significantly. In each transformed line one component of the pattern was
identified as the autophosphorylated kinase (14-19) coded by the
involved oncogene, on the basis of the Mr and, when possible, by
immunoprecipitation by specific antisera (e.g. pp60*src* and p120*abl*).
In addition, P-TYR proteins with different molecular weight were
identified. The molecular weight of the major component of the pattern corresponded to those of proteins decorated by P-TYR antibodies
in immunoblotting.

In mouse fibroblasts transformed by the SR-D strain of RSV, a major
phosphorylated component of 130 Kd, and two closely migrating proteins of 70-65 Kd,were identified. Two minor bands of 85 Kd and 60
Kd were also constantly observed (7). The latter partially over-
lapped with the more radioactive faster component of the 70-65 Kd
doublet. Only the 130 Kd and the 70 Kd components were precipitated
from mouse fibroblasts transformed by the B77 strain of RSV. More-

over, the overall amount of radioactive phosphate transferred to
these proteins was significantly lower than that transferred to the
bands of identical electrophoretic mobility, precipitated from
SR-D transformed cells.

A complete pattern was observed by analyzing the phosphoproteins
precipitated from ^{32}P-γ- ATP labelled detergent-insoluble cell
fraction prepared from rat fibroblasts transformed by RSV either
of the SR-A strain or the tsLA24 PR-A mutant.

The specificity of the kinase reaction was proved by the fact that
no radiolabelled proteins were precipitated from mouse 3T3 or Rat-1
control fibroblast detergent-insoluble preparations, lacking v-*onc*
coded kinases.

In mouse cell lines the 130 and 70-65 Kd proteins, precipitable by
P-TYR antibodies, were the major acceptors of phosphate in the
pp60*src*-catalyzed kinase reaction *in vitro*, being also the major
phosphorylated proteins in the whole detergent-insoluble fraction
examined in PAGE before immunoprecipitation. The radiolabelled
phosphate transferred to these proteins was resistant to alkali
treatment, ruling out the involvement of serine residues as phosphate acceptor sites. Direct evidence for phosphorylation at tyrosine
residue(s) was provided by phosphoaminoacid analysis. After acid
hydrolysis and two dimensional separation, by high voltage electrophoresis and chromatography (18), the radiolabelled phosphate
incorporated in the proteins precipitated by P-TYR antibodies was
found to comigrate with authentic phosphotyrosine.

Phosphopeptide mapping of proteins precipitated by P-TYR antibodies
from cells of different species and/or transformed by different
v-*onc* showed that the two proteins of 130 and 70 Kd observed in

both v-*src* transformed mouse and rat fibroblasts were reciprocally unrelated and unrelated also to the pp60*src* itself. Moreover, no phosphopeptide correlation was found between the proteins of identical Mr (70 Kd) precipitated from cells transformed by different oncogenes (Ferracini, Di Renzo, Naldini and Comoglio, in preparation).

In order to identify tentatively the phosphotyrosine containing proteins precipitated by P-TYR antibodies, their electrophoretic mobility in SDS-PAGE was compared to that of vinculin (Mr 130 Kd) and pp60*src* (Mr 60 Kd), precipitated by specific antisera (7).

The tyrosine-phosphorylated 130 Kd protein, immunoprecipitated by P-TYR antibodies from mouse fibroblast detergent insoluble preparations, comigrated with vinculin immunoprecipitated by specific antisera from ^{35}S-methionine metabolically labelled mouse fibroblasts. However, the 130 Kd protein and vinculin represent different protein species since an antiserum able to precipitate mouse vinculin did not precipitate any tyrosine-phosphorylated protein of 130 Kd from ^{32}P-γ-ATP labelled detergent insoluble fractions of RSV transformed mouse fibroblasts. Control experiments - performed on the same cells metabolically labelled with ^{35}S-methionine - showed that vinculin was indeed precipitated by the same antiserum not only from lysates of whole cells but also from their detergent insoluble fractions thus providing evidence that vinculin was herewith retained (7).

The electrophoretic mobility of the proteins precipitated by P-TYR antibodies was also compared with that of *in vitro* phosphorylated proteins immunoprecipitated by a tumor bearing rabbit serum (TBR). The pp60src comigrated with the minor phosphorylated 60 Kd component.

IMMUNOLOCALIZATION

The localization of P-TYR antibody binding sites was studied by
immunofluorescence and by immunoelectron microscopy in the same
detergent-insoluble preparation of RSV transformed cells used for
precipitation experiments. Detergent-insoluble preparations of
normal and RSV transformed fibroblasts were compared after indirect
decoration with immunogold complexes and analysed in the electron
microscope. While no detectable phosphotyrosine-containing sites
were present in normal cells, in RSV transformed cells gold particles were mostly bound to electron dense granular material associated with the filaments. ATP phosphorylation carried out *in situ*
in the same conditions described in the preceding paragraph yielded
a significant increase in cytoskeleton-associated gold particles
only in transformed cells. Overall, these results are consistent
with the idea that pp60*src* itself and, presumably, some other phosphotyrosine-containing proteins remain tightly attached to detergent-insoluble fractions but do not provide any indication of their
possible location in intact cells (6).

In order to gain information about the distribution of P-TYR proteins in intact cells, mouse, rat and avian fibroblasts transformed
by different retroviruses coding for tyrosine kinases and their
control, non-transformed counterparts were studied by immunofluorescence and P-TYR antibodies after fixation and membrane permeabili-
zation (20). All such transformed cells showed diffuse staining of
the cytoplasm and of the nucleus. While cytoplasmic fluorescence
was specifically observed only in transformed cells, nuclear fluorescence was also displayed by control untransformed fibroblasts.

The distribution of P-TYR containing structures was studied in detail in RSV transformed mouse and duck fibroblasts. In these cells,
well defined fluorescent patches decorated by P-TYR antibodies
were located at the cell periphery, in areas corresponding to the

cytoplasmic side of the plasma membrane. As described in detail
elsewhere (7,20), evidence that significant amounts of P-TYR-proteins were associated with adhesion plaques was based essentially
on the following facts: (a) Simultaneous double labelling, with
fluorescent phalloidin (F-PHD) and P-TYR antibodies, showed elon-
gated streak-like structures located at the endings of residual
stress fibres. (b) Such intensely fluorescent structures corresponded in IRM to well defined dark areas which are known to represent patches of the cell ventral membrane lying less than 10 nm
from the substratum. (c) The distribution of P-TYR antibodies
cross-reacting sites corresponded in size, location and general
morphological features to areas which were also stained by vinculin
antibodies and appeared intensely dark in IRM.

In addition to the location of P-TYR-proteins at cell-substratum
contacts, immunostaining with P-TYR antibodies showed also intense
fluorescence at sites of contact between individual cells *in vitro*.
This finding was noted in most lines of RSV transformed mouse fibroblasts when grown to subconfluency but was never observed in non-
transformed cells.

In some v-*onc* transformed cells, notably in those carrying *src* and
fps, culturing on fibronectin-coated dishes causes the constant
appearance of dot-shaped adhesion sites which correspond to short
protrusions of the ventral surface (6). These structures are markedly different from adhesion plaques not only for their different
morphology but also for a number of other physiological properties
including the fact that they are rapidly formed also in the absence
of serum. These sites, which had been previously observed but not
fully characterized (e.g. 23,25,31), contain a meshwork of microfilament bundles as well as some proteins involved in the architecture of all adhesion structures (6). These dot-shaped sites, denoted
"podosomes" in view of their similarity with cellular feet, accu-

mulate phosphotyrosine-containing proteins (6). Whether the association of the specific kinase with these sites (23) brings about the
reorganization of the adhesive system of transformed cells attached
to fibronectin remains to be established. So far, this finding
strengthens the concept that proteins which have been phosphorylated in tyrosine may be primarily involved in adhesion control and
in the cell structural rearrangement giving rise to the transformed
phenotype (6).

CONCLUSIONS

The transformed phenotype is due to the cumulative effect of morphological and biochemical alterations, including a deep rearrangement of cytoskeleton and adhesion structures (21-23). In cells
transformed by some retroviruses events leading to such changes are
controlled by a single transforming gene (24), whose product is
provided with tyrosine-specific kinase activity (17-19). Although
multiple independent targets were found to be involved in transformation (30-31) and numerous putative substrates of tyrosine kinases
were identified (32-36), it is still debated whether and how kinase
activity affects membrane and cytoskeletal structures responsible
for the control of cell adhesion and morphology.

Several tyrosine kinases, such as pp60src, p120abl and p140fps,
have been reported to be located at the cytoplasmic face of the
plasma membrane (10,12,13). Recently, also one of the putative
pp60src substrates, p36-39, has been found to be located at the
cytoplasmic surface of the plasma membrane (37-40).

P-TYR antibodies have been shown to cross-react specifically with
phosphotyrosine and to identify P-TYR proteins (3,41); previous
work has shown that P-TYR proteins may be precipitated by antibodies from detergent insoluble cell preparations of RSV trans-

formed fibroblasts, containing cytoskeleton components and plasma membrane domains (7). We have also shown that P-TYR antibodies intensely decorate restricted areas of the plasma membrane in RSV-transformed cells, giving direct evidence that the pp60src induced transformation is accompanied by increased phosphorylation of proteins constitutive or associated with adhesion plaques and cell-cell contacts.

Moreover, the data reviewed in this paper show that protein kinases coded by v-*src*, v-*abl*, v-*fes* of ST-FeSV and, to a lesser extent, those coded by v-*fps* and v-*fes* of GA-FeSV, phosphorylate at tyrosine detergent insoluble molecules. The different degree of phosphorylation reflected the different association of the respective transforming proteins with the detergent insoluble cell fractions (10,12,13).

These data also suggest that in the same cells transformed by different retroviruses, different phosphoproteins are found. Evidence of the non-identity between these phosphoproteins and the involved v-*onc* coded kinases has already been obtained in the case of v-*src* transformed fibroblasts by phosphopeptide fingerprinting (Ferracini, Di Renzo, Naldini and Comoglio, in preparation). Since it has been shown that tyrosine kinases phosphorylate *in vitro* any phosphate acceptor protein brought in physical contact (e.g. immunoglobulins, 26), the observed target specificity seems to originate from the localization of tyrosine kinases in a specific microenvironment, such as that provided by the association with detergent insoluble macromolecular complexes. In this respect, it has already been demonstrated that the subcellular localization of pp60 *src*, p120 *abl* and P130 *fps* tyrosine kinases is a specific property (10,12,13,29), depending on structural features of the molecules, among which the presence of a fatty acid residue covalently bound to the N-terminal protein domain (27,28).

ACKNOWLEDGEMENTS

The skilful technical assistance of M.R. Amedeo and P. Rossino is
gratefully acknowledged. The authors thank Prof. Th. Wieland for
providing fluorescein labelled phalloidin. This work was supported
by the Italian National Research Council (C.N.R.), Progetto Fina-
lizzato "Oncologia".

REFERENCES

1) BISHOP, J.M. (1983). Cellular oncogenes and retroviruses. Ann.
   Rev. Biochem., 52, 301-316.

2) ROSS, A.H., BALTIMORE, D. & EISEN, H.N. (1981). Phosphotyrosine
   containing proteins isolated by affinity chromatography with
   antibodies to a synthetic hapten. Nature, 294, 654-656.

3) TABACHNICK, M. & SOBOTKA, H. (1960). A spectrophotometric study
   of the coupling of diazotized arsanilic acid with proteins. J.
   Biol. Chem., 235, 1051-1054.

4) COMOGLIO, P.M., DI RENZO, M.F., NALDINI, L. & MARCHISIO, P.C.
   (1984). Identification of oncogene coded kinase cellular targets by phosphotyrosine antibodies. In: "Recent Advances in
   Tumor Immunology: From oncogenes to tumor antigens". G. Giraldo,
   ed., Elsevier, Amsterdam.

5) PRAT, M. & COMOGLIO, P.M. (1976). A solid-state competitive
   binding radioimmunoassay for measurement of antigens solubilized from membranes. J. Immunol. Methods, 9, 267-272.

6) TARONE, G., CIRILLO, D., GIANCOTTI, F.G., COMOGLIO, P.M. &
   MARCHISIO, P.C. (1984). Rous sarcoma virus transformed fibroblasts adhere primarily at discrete protrusions of the ventral
   membrane called podosomes. Exp. Cell Res., submitted for publication.

7) COMOGLIO, P.M., DI RENZO, M.F., TARONE, G., GIANCOTTI, F.G.,
   NALDINI, L. & MARCHISIO, P.C. (1984). Detection of phosphotyrosine-containing proteins in the detergent-insoluble frac-

tion of RSV-transformed fibroblasts by azobenzene phosphonate antibodies. EMBO J., 3, 483-487.

8) TOWBIN, H., STAEHELIN, T. & GORDON, J. (1979). Electrophoretic transfer of proteins from polyacrylamide gels to nitrocellulose sheets: Procedure and some applications. Proc. Natl. Acad. Sci. USA, 76, 4350-4354.

9) BURNETTE, W.N. (1981). "Western blotting": Electrophoretic transfer of proteins from sodium dodecyl sulfate-polyacrylamide gels to unmodified nitrocellulose and radiographic detection with antibody and radioiodined Protein A. Anal. Biochem., 112, 195-203.

10) BURR, J., DREYFUSS, G., PENMAN, S. & BUCHANAN, J. (1980). Association of the src gene product of Rous sarcoma virus with cytoskeletal structure of chicken embryo fibroblasts. Proc. Natl. Acad. Sci. USA, 77, 3484-3488.

11) GACON, G., GISSELBRECHT, S., PIAU, J.P., FISZMAN, M.Y. & FISHER, S. (1982). Phosphorylation of the subcellular matrix in cells tranformed by Rous sarcoma virus. Eur. J. Biochem., 125, 453-456.

12) BOSS, M.A., DREYFUSS, G. & BALTIMORE, D. (1981). Localization of the Abelson murine leukemia virus protein in a detergent insoluble subcellular matrix: architecture of the protein. J. Virol., 40, 472-479.

13) FELDMAN, R.A., WANG, E. & HANAFUSA, H. (1983). Cytoplasmic localization of the transforming protein of Fujinami sarcoma virus: salt sensitive association with subcellular components. J. Virol. 45, 782-789.

14) WITTE, O.N., DASGUPTA, A. & BALTIMORE, D. (1980). Abelson murine leukemia virus protein is phosphorylated *in vitro* to form phosphotyrosine. Nature, 283, 826-832.

15) FELDMAN, R.A., HANAFUSA, T. & HANAFUSA, H. (1980). Characterization of protein kinase activity associated with the transforming gene product of Fujinami sarcoma virus. Cell, 22, 757-765.

16) VAN DE VEN, W.J.M., REYNOLDS, F.H. & STEPHENSON, J.R. (1980). The non structural components of polyproteins encoded by replication defective mammalian transforming retroviruses are phosphorylated and have associated protein kinase activity. Virology, 101, 185-197.

17) COLLETT, M.S., PURCHIO, A.F. & ERIKSON, R.L. (1980). Avian sarcoma virus-transforming protein pp60src shows protein kinase activity specific for tyrosine. Nature, 285, 167-168.

18) HUNTER, T. & SEFTON, B. (1980). Transforming gene product of Rous sarcoma virus phosphorylates tyrosine. Proc. Natl. Acad. Sci. USA, 77, 1311-1315.

19) LEVINSON, A.D., OPPERMANN, H., LEVINTOW, L., VARMUS, H.E. & BISHOP, J.M. (1978). Evidence that the transforming gene of avian sarcoma virus encodes a protein kinase associated with a phosphoprotein. Cell, 15, 561-572.

20) MARCHISIO, P.C., DI RENZO, M.F. & COMOGLIO, P.M. (1984). Immunofluorescence localization of phosphotyrosine containing proteins in RSV-transformed mouse fibroblasts. Exp. Cell Res. 154, 112-124.

21) EDELMAN, G.M. & YAHARA, I. (1976). Temperature sensitive changes in surface modulating assemblies of fibroblasts transformed by mutants of Rous sarcoma virus. Proc. Natl. Acad. Sci. USA, 73, 2047-2051.

22) WANG, E. & GOLDBERG, A.R. (1976). Changes in microfilament organization and surface topography upon transformation of chick embryo fibroblasts with Rous sarcoma virus. Proc. Natl. Acad. Sci. USA, 73, 4065-4069.

23) DAVID-PFEUTY, T. & SINGER, S.J. (1980). Altered distribution of cytoskeletal proteins vinculin and alpha-actinin in cultured fibroblasts transformed by Rous sarcoma virus. Proc. Natl. Acad. Sci. USA, 77, 6687-6691.

24) HANAFUSA, H. (1977). Cell transformation by RNA retroviruses. In: "Comprehensive Virology", H. Fraenkel-Conrat & R.R. Wagner, eds., Plenum Press, New York, 10, 401-419.

25) WANG, E., YIN, H.E., KRUEGER, J.G., CALIGURI, L.A. & TAMM, I. (1984). Unphosphorylated gelsolin is localized in regions of cell-substratum contact or attachment in Rous sarcoma virus transformed rat cells. J. Cell Biol., 98, 761-771.

26) ERIKSON, R.L., COLLETT, M.S., ERIKSON, E. & PURCHIO, A.F. (1979). Evidence that the avian sarcoma virus transformed gene product is a cyclic AMP-independent protein kinase. Proc. Natl. Acad. Sci. USA, 76, 6260-6268.

27) KRUEGER, J.G., GARBER, E.A., GOLDBERG, A.R. & HANAFUSA, H. (1982). Changes in amino-terminal sequences of pp60src leak to decreased membrane association and decreased *in vivo* tumorigenicity. Cell, 28, 889-895.

28) SEFTON, B.M., TROWBRIDGE, I.S., COOPER, J. & SCOLNICK, J.A. (1982). The transforming protein of Rous sarcoma virus, Harvey sarcoma virus and Abelson virus contain tightly bound lipid. Cell, 31, 465-474.

29) LEVINSON, D.A., COURTNEIDGE, A. & BISHOP, M. (1981). Structural and functional domains of the RSV transforming protein (pp60src). Proc. Natl. Acad. Sci. USA, 78, 1624-1628.

30) BEUG, H., CLAVIEZ, M., JOCKUSCH, B.M. & GRAFT, T. (1978). Differential expression of Rous sarcoma virus specific transformation parameters in enucleated cells. Cell, 14, 843-856.

31) ROHRSCHNEIDER, L.R. & ROSOK, M.J. (1983). Transformation parameters and pp60src localization in cells infected with partial transformation mutants of Rous sarcoma virus. Mol. Cell. Biol. 3, 731-742.

32) RADKE, K. & MARTIN, G.S. (1979). Transformation by Rous sarcoma virus: effects of src gene expression on the synthesis and phosphorylation of cellular polypeptides. Proc. Natl. Acad. Sci. USA, 76, 5212-5216.

33) ERIKSON, E. & ERIKSON, R.L. (1980). Identification of a cellular protein substrate phosphorylated by the avian sarcoma

virus-transforming gene product. Cell, 21, 829-836.

34) COOPER, J. & HUNTER, T. (1981). Changes in protein phosphorylation in Rous sarcoma virus-transformed chicken embryo cells.
Mol. Cell. Biol., 1, 165-171.

35) COOPER, J. & HUNTER, T. (1982). Discrete primary locations of
a tyrosine protein kinase and of three proteins that contain
phosphotyrosine in virally-transformed chick fibroblasts. J.
Cell Biol., 94, 287-296.

36) COOPER, J. & HUNTER, T. (1983). Identification and characterization of cellular targets for tyrosine protein kinase. J.
Biol. Chem., 258, 1108-1119.

37) GREENBERG, M. & EDELMAN, G.M. (1983). The 34Kd pp60src substrate is located at the inner face of the plasma membrane.
Cell, 33, 767-779.

38) NIGG, E.A., COOPER, J.A. & HUNTER, T. (1983). Immunofluores-
cent localization of a 39,000 dalton substrate of tyrosine
protein kinases to the cytoplasmic surface of the plasma membrane. J. Cell Biol., 96, 1601-1608.

39) RADKE, K., CARTER, C., MOSS, P., DEHAZYA, P., SCHLIWA, M. &
MARTIN, G.S. (1983). Membrane association of a 36,000 d substrate for tyrosine phosphorylation in chicken fibroblasts
transformed by avian sarcoma virus. J. Cell Biol. 97, 1601-1608.

40) COURTNEIDGE, S., RALSTON, R., ALITALO, K. & BISHOP, M.J.
(1983). Subcellular location of an abundant substrate (p36)
for tyrosine specific kinase. Mol. Cell. Biol. 3, 340-350.

41) FRACKLETON, A.R., ROSS, A. & EISEN, H.N. (1983). Characterization and use of monoclonal antibodies for isolation of phos-
photyrosyl proteins from retrovirus transformed cells and
growth factor-stimulated cells. Mol. Cell. Biol.,3, 1343-1348.

THE TRANSFORMATION CAPACITY OF EARLY SV40 DNA FRAGMENTS

Adolf Graessmann & Monika Graessmann

Institut fuer Molekularbiologie und Biochemie
Freie Universitaet Berlin
D-1000 Berlin 33

INTRODUCTION

Transformation of primary tissue culture cells or cells of established lines is efficiently induced by microinjection of intact SV40 DNA or subgenomic DNA fragments which contain only the early genome region (1,2). This SV40 part codes for two related proteins, the large T-(tumor) antigen and the small t-antigen. Transformation experiments with temperature sensitive (tsA) - and small t-antigen negative mutants have shown that synthesis of T-antigen is required and sufficient for induction and maintenance of the transformed state (for review see Tooze, 3). In the meantime many details about the T-antigen have become known, but it is still unclear which of the T-antigen specific function(s) is crucial for the SV40 oncogenicity.

This matter is complicated by the fact that the large T-antigen is a multifunctional protein with different biological and biochemical functions (3). To further analyse this question we tested the biological activity and transformation capacity of different early SV40 DNA fragments following microinjection into cultured cells. These experiments have shown that a DNA fragment which codes only for the second large T-antigen exon has the capability to fully transform rat embryo fibroblast cells.

THE BIOLOGICAL ACTIVITY OF THE SECOND T-ANTIGEN EXON

To analyse the biological activity of the second T-antigen exon, SV40 DNA I was cleaved with the TaqI and the Bam HI endonucleases (Fig. 1). After preparation and ligation of Sal linkers, this DNA fragment was inserted into the Sal site of the pBR 322 DNA and cloned in *E. coli*. After reisolation, the purified Taq/Bam SV40 DNA fragment was microinjected into the nuclei of TC7 (monkey) and Ref 52 (rat) cells. As expected, this promoter free DNA segment was biologically inactive. None of the recipient cells stained positive for T-antigen at any time point after microinjection.

However T-antigen positive cells were obtained after microinjection of RNA complementary to the Taq/Bam DNA. This RNA was obtained by *in vitro* transcription of the SV40 DNA segment with the *E. coli* RNA polymerase (4). T-antigen positive nuclei were first demonstrable 2-3 hours after cRNA injection and about 50% of the injected cells stained positive 20-24 hours later (Fig. 2). The cytoplasmic fluorescence did not exceed the intensity of mock injected cells. These experiments have shown that the second exon RNA can be translated *in vivo* although this cRNA does not have the authentic 5'end of the early SV40 mRNA. To test further if the second large T-antigen exon per se is biologically active, two T-antigen specific functions were analysed, namely stimulation of cell DNA synthesis and helper function for Adeno 2 virus (AD_2) in monkey cells (5). Stimulation of cell DNA synthesis was tested in confluent cultures of primary mouse kidney cells. After cRNA

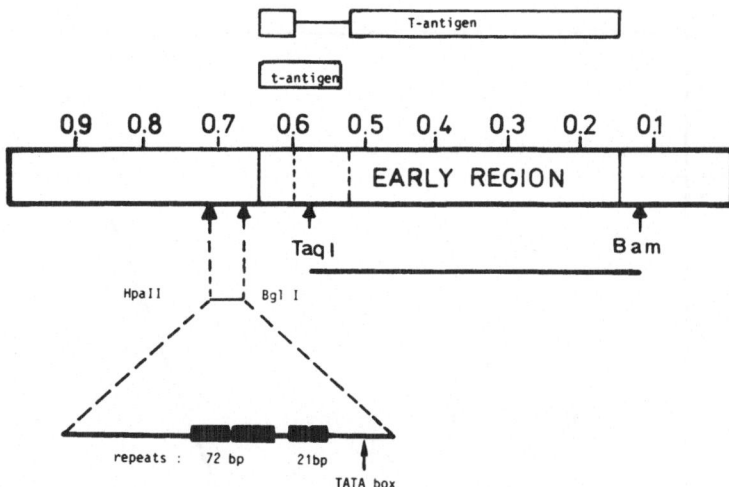

Fig. 1. *Assignment of the HpaII/BglI and TaqI/Bam HI DNA fragments to the physical map of the SV40 genome.* On the left bottom part of the figure an amplification of the HpaII/BglI fragment with the positions of the 72 bp and 21 bp repeats and the TATA box. The bars at the top indicate the positions of the small t-antigen and the two large T-antigen exons.

injection, ^{3}H-thymidine (final concentration 0.1 µCi/ml) was added to the culture medium for 20 hours. Thereafter cells were fixed and stained for T-antigen and processed for autoradiography. These experiments have shown that the Taq/Bam cRNA has the capability to induce cell DNA synthesis. About 80% of T-antigen positive cells incorporated thymidine (mock injected cells only 5%). The Taq/Bam cRNA also mediates as a helper for AD_2 in monkey cells. TC7 cells, preinjected with the cRNA and infected with the virus (5-10 pfu/cell), synthesized the adeno fiber protein as tested by immunofluorescence staining. Synthesis of this protein is an indirect marker for AD_2 DNA replication and hence for helper activity.

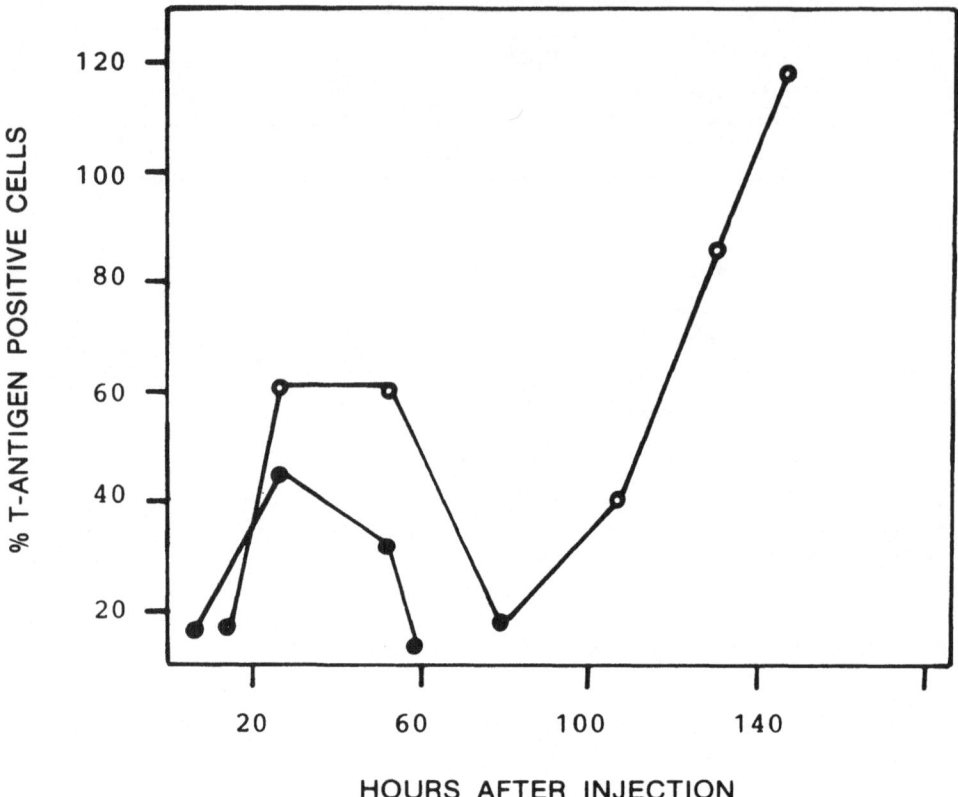

Fig. 2. *Time course of T-antigen synthesis in Ref 52 cells after microinjection of :* ●————● *Taq/Bam HI specific cRNA;* o————o *HpaII/Bg1I and TaqI/BamHI fragment mixture.*

TRANSFORMATION CAPACITY OF THE SECOND T-ANTIGEN EXON

Since maintenance of the transformed state requires permanent T-antigen synthesis (6,7), we asked whether expression of the Taq/Bam HI DNA fragment can be obtained by coinjection with the early HpaII/Bg1I DNA segment. As shown in Fig. 1 this DNA fragment contains the early SV40 promoter region, the 72 bp enhancer repeats and the TATA box (8). The HpaII/Bg1I fragment was also cloned in

the SalI site of pBR 322. Without prior *in vitro* ligation the isolated TaqI/BamHI and the HpaII/Bg1I fragments were microinjected
at a concentration of 0.1 mg/ml each into the nuclei of Ref 52
cells. The time course of T-antigen synthesis in these cells is
shown in Fig. 2. About 60% of the injected Ref 52 cells exhibited
a strong intranuclear T-antigen fluorescence 20-40 hours after injection. The decline of T-antigen positive cells around 60 hours
after injection was followed again by an increase in the number
of positive cells 80-120 hours later (Fig. 2). These T-antigen
positive cells appeared in a clonal fashion. To obtain T-antigen
positive cell lines, 300 Ref 52 cells were microinjected in groups
of 25 cells with the DNA fragment mixture.

Ref 52 cells were chosen for these experiments, as Ref 52 is an
exceptionally untransformed established rat line, which has not
shown any spontaneous colony formation in soft agar so far. The
recipient cells were grown on glass slides subdivided into num-
bered squares of 2 mm^2 size. One week after injection cells within these squares were isolated and cultivated further in micro-
wells (8). After the second passage in 60 mm dishes,control slides
with the cells were stained for T-antigen. From the 12 independent
isolates obtained, 7 contained T-antigen positive cells (Ref-Taq/
Bam line 1-7). Further enrichment of T-antigen positive cells was
obtained by mild trypsinisation and serial dilution experiments.
Between passage number 5-8 the lines contained on the average 90-
95% T-antigen positive cells. Selective pressure e.g. low serum
concentration was not applied.

To characterize the Taq/Bam T-antigen polypeptides synthesized in
these lines, cells were labelled with [^{35}S]-methionine for two
hours, and the immunoprecipitates were analysed by discontinuous
SDS polyacrylamide gel electrophoresis (7-15%). According to the

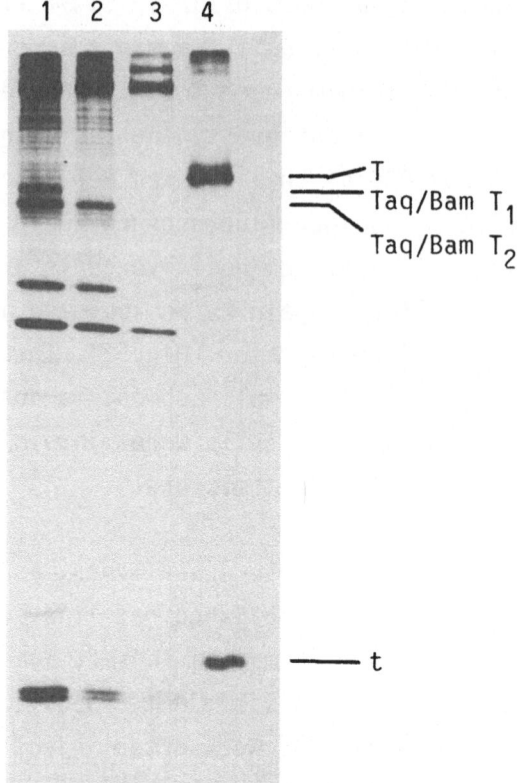

Fig. 3. *Fluorogram of a SDS polyacrylamide gel (7-15%) of proteins immunoprecipitated with T-serum from the following cells:* Lane 1: Ref-Taq/Bam line 5; Lane 2: Ref-Taq/Bam line 7; Lane 3: Ref 52; Lane 4: Cos 7.

T-antigen polypeptides synthesized in these cells two categories of cell lines were obtained. Cells of the Ref-Taq/Bam lines 1-5 synthesized two T-antigen related polypeptides (Taq/Bam T_1 and Taq/Bam T_2) not demonstrable in the Ref-wt cells (Fig. 3). Cells of the Ref-Taq/Bam line 6 and 7 synthesized only the lower molecular weight Taq/Bam T_2 form (Fig. 3). Both Taq/Bam polypeptides are also precipitable by a serum which is specific for the last 11 amino acids of the C-terminus of the T-antigen (serum kindly

provided by G. Walter, 9). This indicates that the two truncated
T-antigen molecules differ in their amino terminal portion.

A large range of different tumor cell types contains the cellular
p53 protein at elevated concentrations (10). Enhanced concentrations of this protein is also demonstrable in all Ref Taq/Bam
lines both by immunofluorescence staining and immunoprecipitation.
To analyse the state of the Taq/Bam and the HpaII/Bg1I DNA fragments in the transformed cells, cellular DNA was extracted from
the Ref-Taq/Bam line 5 and the Ref-Taq/Bam line 7 cells. Following
digestion of the DNA with different restriction endonucleases two
blots were prepared. One was hybridized with the nick-translated
Taq/Bam DNA fragment and the other with the nick-translated HpaII/
Bg1I SV40 fragment. Fig. 4 shows the blot of the Taq/Bam line 5
DNA hybridized with the nick-translated Taq/Bam DNA fragment. Lane
1 contains the cellular DNA digested with the Bg1II and BamHI
endonuclease. These enzymes do not cut the SV40 DNA fragments (the
cloned Taq/Bam DNA fragment does not contain the intact BamHI site).
Six independent bands are demonstrable. The blot hybridized with
the HpaII/Bg1I fragment (Fig. 4b) contains five HpaII/Bg1I specific
bands in lane 1, all at the same position as the Taq/Bam DNA.
Direct cointegration of both types of SV40 DNA fragments was demonstrated after further digestion of the cellular DNA with the
SalI enzyme. This treatment separated most of the common integra-
tions and generated SV40 DNA fragments of the original size (lane
2). The Taq/Bam and HpaII/Bg1I DNA fragments could also be rescued
from the cellular DNA of the Taq-Bam line 7 by Sal I treatment
(data not shown).

To analyse the transformation capacity of the Taq/Bam DNA fragment
we monitored the morphology and anchorage-independent growth of
the Taq/Bam cell lines 1-7 for a period of about 1.5 years. Lines
with about 95% T-antigen positive cells were obtained between

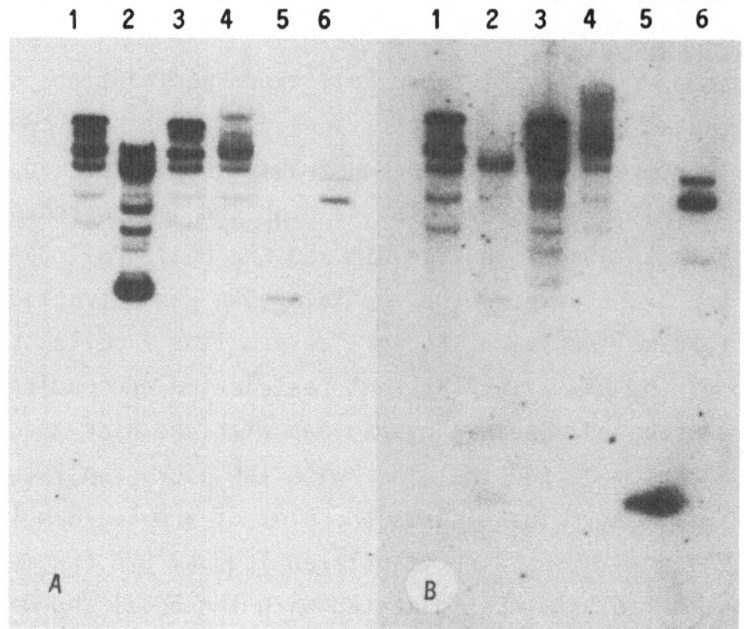

Fig. 4. *Southern blotting analysis of the Ref-Taq/Bam line 5.* Following extraction of the cellular DNA and digestion with different restriction endonucleases, one blot was prepared and divided for hybridization in two parts. Part A was hybridized with nick-translated ^{32}P-Taq/Bam SV40 DNA fragment, and part B with ^{32}P-HpaII/Bg1I SV40 DNA fragment. The lanes 1-4 contain 8 µg of high molecular weight DNA restricted with the endonucleases: 1. Bg1II and BamHI; 2. Bg1II and SalI; 3. Bg1II and EcoRI; 4. Bg1II. Lane 5 contains a mixture of the SV40 DNA fragments HpaII/Bg1I and Taq-Bam, and lane 6 marker DNAs I, II and III.

passage numbers 5-10. At this point of time the morphology of the cells was close to the parental Ref 52 cells. With continuous further cultivation the shape of the line 1-5 cells changed gradually from normal to the morphology of fully SV40 transformed rat cells. The cells become smaller, exhibited a reduced cytoplasmic actin cable network, a change of the pericellular matrix, an increase of plasminogen activator secretion, and a low serum requirement but anchorage, independent growth did not directly correlate with these alterations. At passage numbers 20-25 when the morphology of the Taq-Bam cell lines 1-5 was close to the morphology of the Ref-wt-cells only 0.1-0.2% of these cells formed colonies in soft agar. At passage numbers 50-60 the rate was still only 1-2%. This observation indicates that under standard culture conditions the cells with anchorage growth characteristics do not have a significant growth advantage over the rest of the cell population. Furthermore, it was not possible to select these cells by other means than soft agar (e.g. low serum concentration, mild trypsination). However, once isolated from soft agar, these cells exhibited a high efficiency in colony formation (60-80%) after replating into semi-solid media (agar or agarose).

With increases passage numbers the cells of the Ref Taq/Bam lines 6 and 7 did not change their cell morphology and growth behaviour significantly. As shown in Fig. 5, the morphology of these cells was still close to the morphology of the parental Ref 52 cells, and the cells still contained their well developed actin cable structure after a continuous culture period of 1.5 years. From the transformation criteria tested, the line 6 and 7 cells only exhibited a low serum requirement, but growth in soft agar or in agarose was not demonstrable.

DISCUSSION

Here it was shown that the early SV40 Taq/Bam DNA fragment (m.p. 0.566-0.144) has the capability to fully transform the Ref 52 cells. Expression of this promoter minus SV40 DNA segment was possible by coinjection with the early SV40 promoter/enhancer HpaII/Bg1I DNA fragment (m.p. 0.72-0.66) (8,11). *In vitro* ligation of the two fragments was not required because coligation occurred

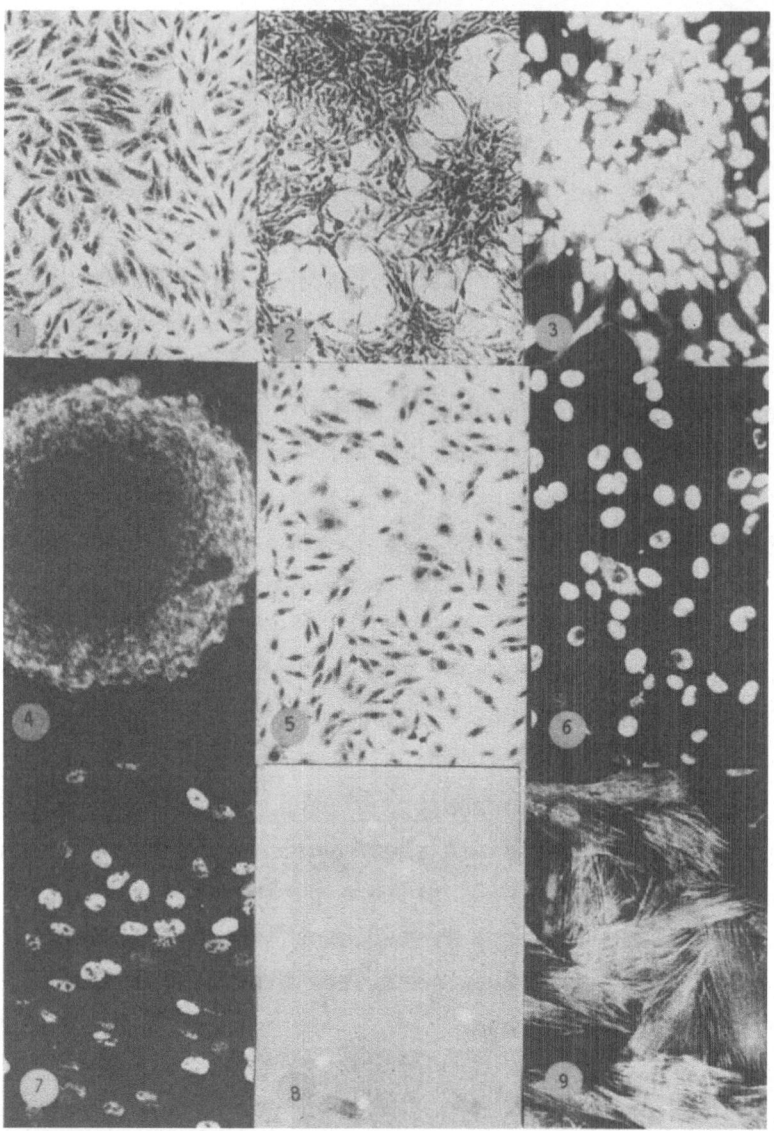

Fig. 5. *Morphology of the Ref 52 cells and Ref-Taq/Bam transfor-mants.* 1) Phase contrast picture of confluent Ref 52 cells. 2) Ref-Taq/Bam line 5. 3) T-antigen positive Ref-Taq/Bam line 5 cells. 4) A clone of the Ref-Taq/Bam line 5, two weeks after plating in soft agar. 5) The Ref-Taq/Bam line 7 cells. 6) T-antigen positive Ref-Taq/Bam line 7 cells. 7) Ref-Taq/Bam line 7 cells stained with anti p53 serum. 8) Ref-Taq/Bam line 7 cells, two weeks after plating into soft agar. 9) Ref-Taq/Bam line 7 cells stained with anti actin serum.

inside of the recipient cells. DNA reextraction experiments and
Southern blot analysis have shown that formation of polymers by
end to end ligation started as early as 1-2 minutes after injection. This oligomerisation process was certainly facilitated by
the fact that the DNA fragments had the same overlapping ends
(Sal linker). However, fragments with different or blunt ends were
also coligated, although with a lower efficiency. Stable T-antigen
positive cell lines were isolated with a frequency of 1-5%. This
remarkably high transformation rate was only observed after intra-
nuclear but not after cytoplasmic injection of the fragment mixture. Rapid degradation of the DNA fragments inside of the cytoplasm is at least partly responsible for this lower efficiency
(Graessmann et $al.$, in preparation). The Ref Taq/Bam lines fall
into two distinct categories of cells, namely into maximally
and into minimally transformed cells. The maximal transformants
show among other things anchorage-independent growth low serum
requirement, loss of actin cable structure and alteration of the
perinuclear matrix (fibronectine). The only transformation parameter attributable to the minimal transformants (Ref-Taq/Bam
lines 6 and 7) is the reduced serum requirement. Cells of the
first category synthesize two T-antigen exon specific polypeptides (Taq/Bam T1 and T2). Up to now, we have not tested, if
translation of the Taq/Bam mRNA or cRNA starts with AUG. If so,
the triplets coding for the T-antigen amino acids 109 and 176
are the potential initiation signals. These are the first two
AUGs of the second T-antigen exon in the reading frame of the
wild type protein (Fig. 6).

The two minimally transformed Taq/Bam cell lines 6 and 7 synthesize only the Taq/Bam T2 polypeptide. This observation implies
that the amino acid sequence 82 to 176 of large T-antigen is important for maximal cell transformation (see Fig. 6). It was

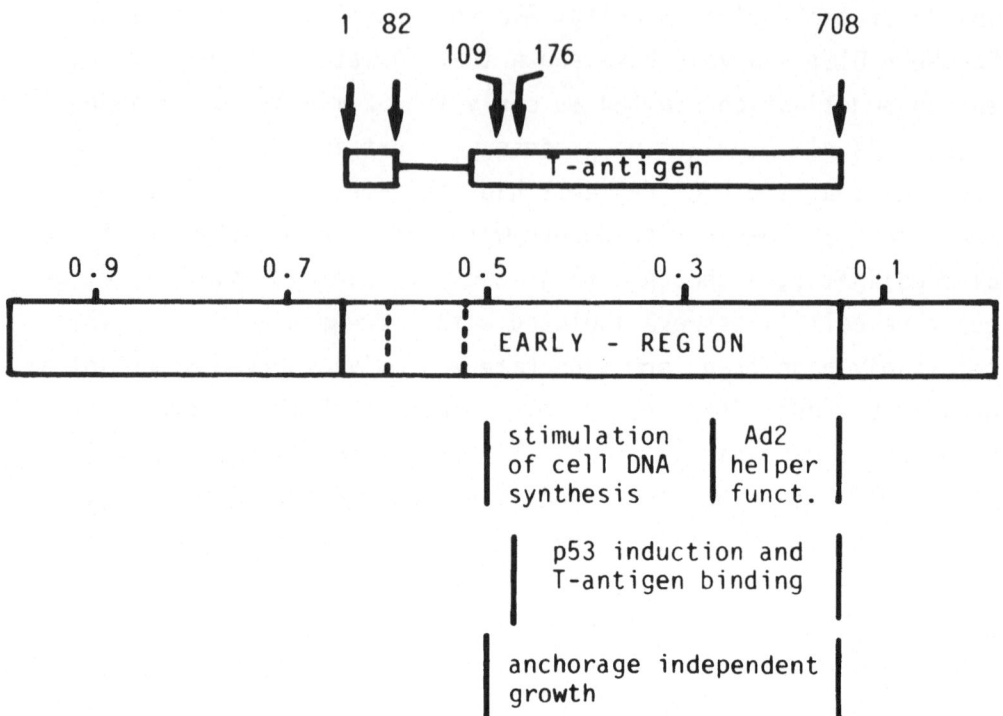

Fig. 6. *Functional map of the second T-antigen exon.* Arrows with the number 109 and 176 indicate the positions of the first two methionine of the second T-antigen exon.

shown recently that the amino acid sequence 82-130 is part of the specific DNA binding domaine of the T-antigen (12), however these results cannot yet be considered as proof, that this T-antigen activity is essential for anchorage-independent growth. Up to now it has not even been established that the nuclear localisation of the T-antigen is essential for cell transformation. It is known that small amounts of T-antigen are also associated with cell membrane (13,14).

The DNA fragment injection experiments have shown so far that the first T-antigen exon and small t-antigen are not essential for maximal transformation of Ref 52 cells. Striking however, is the long latency period required for the cells to acquire anchorage-independent growth. There is increasing evidence that maximal cell transformation is a multi step process, and that the SV40 specific oncogenicity is a combined effect of the large and small t-antigen (15,16). In this team effort the small t-antigen accelerates initiation of cell transformation but functions of the second large T-antigen exon are essential for the cells to acquire the fully transformed state. This assumption is in accordance with the observation that small t-antigen negative SV40 deletion mutants (e.g. dl 884) have a lower transformation efficiency than the wild type virus, and a long latency period for manifestation of anchorage-independent growth (17,18).

REFERENCES

1) GRAESSMANN, M., GRAESSMANN, A., & MUELLER, C. (1980). Cold Spring Harbor Symposia on Quantitative Biology, 44, 605.
2) GRAESSMANN, A., GRAESSMANN, M., TOPP, W.C., & BOTCHAN, M. (1979). J. Virol., 32, 989.
3) TOOZE, J. ed. (1980). Cold Spring Harbor Laboratory, Cold Spring Harbor, NY.
4) GRAESSMANN, M. & GRAESSMANN, A. (1976). Proc. Natl. Acad. Sci. USA, 73, 366.
5) TJIAN, R., FEY, G., & GRAESSMANN, A. (1978). Proc. Natl. Acad. Sci. USA, 75, 1279
6) MARTIN, R.G. & CHOU, J.Y. (1975). J. Virol., 15, 599.

7) TEGTMYER, P. (1975). J. Virol., 15, 613.

8) BENOIST, C. & CHAMBON, P. (1981). Nature, 290, 304.

9) WALTER, G., SCHEIDTMANN, K.H., CARBONE, A., LAUDANO, A.P., & DOOLITTLE, R.F. (1980). Proc. Natl. Acad. Sci. USA, 77, 5197.

10) LEVINE, A. (1982). Adv. in Cancer Res. 37, eds G. Klein & S. Weinhause, Academic Press, New York, London, Paris, San Diego, San Francisco, Sao Paulo, Sydney, Tokyo, Toronto.

11) MUELLER, C., GRAESSMANN, A., & GRAESSMANN, M. (1978). Cell, 14, 579.

12) MORRISON, B., KRESS, M., KHOURY, G., & JAY, G. (1983). J. Virol., 47, 106.

13) DEPPERT, W. & HENNING, R. (1979). Cold Spring Harbor Symposia on Quantitative Biology, 44, 225.

14) SOULE, H.R. & BUTEL, J.S. (1979). J. Virol., 30, 523.

15) LAND, H., PARADA, L.F., & WEINBERG, R.A. (1983). Nature, 304, 596.

16) TOPP, W.C. & RIFKIN, D.B. (1980). Virology, 106, 282.

17) SLEIGH, W.J., TOPP, W.C., HANICH, R., & SAMBROOK, J.F. (1978). Cell, 14, 79.

18) RUBIN, H., FIGGE, J., BLADON, M.T., BO CHEN, L., ELLMANN, M., BIKEL, I., FARRELL, M., & LIVINGSTON, D.M. (1982). Cell, 30, 469.

THE TRANSFORMING GENES OF POLYOMA VIRUS

Minoo Rassoulzadegan & Francois Cuzin

Unité de Génétique Moléculaire des Papovavirus (INSERM U273), Centre de Biochimie, Université de Nice, Parc Valrose, 06034, Nice, France

INTRODUCTION

As indicated by its name, polyoma virus induces a variety of histologically different tumors (Stewart, 1960). Like its close relative SV40, the virus is in fact able to transform a large spectrum of differenciated rodent cells. Most commonly used target cells are rodent (rat, mouse, hamster) fibroblasts: two types of cultures are available, the most normal primary cultures derived from 12-15 day rodent embryos (REF cells) and the established (3T3) lines, which, although in principle non tumorigenic, have acquired one of the characters of a transformed cell, the ability to grow in culture for an indefinite number of generations (Todaro & Green, 1963). Transformation of REF or 3T3 cells leads to the same "highly transformed" phenotype: as originally shown by the pioneer work of Vogt & Dulbecco (1962,1963), these cells grow in all cases indefinitely in culture, grow in culture past confluency, in suspension, in the absence of serum factors, exhibit a completely disorganized cytoskeleton, etc. These cell lines are highly tumorigenic in the syngeneic animal.

MORE THAN ONE VIRAL GENE IS INVOLVED IN INDUCING THE TUMORAL PROCESS

Part of the viral genome, corresponding to the early genes of the lytic cycle, is continuously expressed in transformed and tumor cells, from viral DNA integrated into the host chromosomes. The same DNA sequences code for three distinct proteins, by a complex process of alternate splicing. These three proteins (T antigens) are designated respectively large T (105 kDa), middle T (56 kDa) and small T (21 kDa) (see Ito, 1980 for review).

Genetic studies have identified two complementation groups: - the *tsa* mutations (Fried, 1965) do not affect the middle T and small T proteins, but confer to large T an abnormal thermolability, so that the protein is heat-inativated above 37°C (Ito *et al.*, 1977), - the *hrt* mutations (Benjamin, 1970) lead to the loss of both middle T and small T, a normal large T protein being synthesized (Schaffhausen *et al.*, 1978).

The observations that (i) *hrt* mutants are unable to transform (Fluck *et al.*, 1977), and (ii) that *tsa* transformed lines could be isolated at 33°C which reverted to a normal phenotype at 40°C (Seif & Cuzin, 1977, Rassoulzadegan *et al.*, 1981), led us to formulate the hypothesis that the maintenance of a fully transformed state might require at least two distinct gene products, large T and middle or small T (Rassoulzadegan *et al.*, 1980, 1981).

Genetic engineering techniques allowed Kamen and his colleagues to construct modified polyoma genomes which could only express one of the early proteins (Treisman *et al.*, 1981, Tyndall *et al.*, 1981): the three genes thus separated will be designated *plt* (large T protein), *pmt* (middle T protein) and *pst* (small T protein).

Transformation of established (3T3) cells could be achieved by transfer of the *pmt* gene only, provided that the medium contains high enough concentrations of serum (10%) (Treisman *et al.*, 1981). However, these transformed cells (MTT lines) revert to a normal phenotype in the presence of only 0.5% serum (Rassoulzadegan *et al.*, 1982). By contrast, when transfered into primary embryo cells, the *pmt* gene is not able to induce focus formation (Rassoulzadegan *et al.*, 1983). Two functions in cell transformation were thus identified, which are not effected by the middle T protein, cell immortalization and ability to grow in low serum medium.

Transfer of the *plt* gene into serum-dependent established rat cells (FR3T3) leads to the emergence of lines able to grow in the presence of only 0.5% serum. When transferred into primary rat embryo cells, it leads to the appearance of cell lines which are both capable of undefinite growth in culture ("immortalization") and of growth at low serum concentrations. MTT cells, transformed only by the *pmt* gene, can be complemented by *plt* for growth in suspension or focus formation at low serum concentrations and cell lines which were established from rat embryo cells after transfer of *plt* can be subsequently transformed by transfer of *pmt* only. These results clearly established a two-step transformation mechanism: immortalization by *plt* followed by tumorization by *pmt* (Rassoulzadegan *et al.*, 1982, 1983 & Cuzin, 1984 for review).

MULTIGENIC TRANSFORMATION IN OTHER SYSTEMS

The two-step model initially demonstrated for polyoma virus could be extended to other oncogenes and appears now as a general model for the known progressivity of tumoral transformation.

In parallel with our own work on polyoma, that of Van der Eb and his colleagues had demonstrated that two groups of genes, desig-

nated E1A and E1B, were required for cell transformation by adenoviruses (Houweling *et al.*, 1980, Van den Elsen *et al.*, 1982). E1A
was shown to confer immortality by itself. The presence of multiple
oncogenes, without any detectable evolutionary relationship (sequence homology) with host cellular genes, thus appear to distinguish drastically these two groups of DNA tumor viruses, and possibly others (Herpesviruses?) from the genetically more simple Retroviruses.

Land, Parada & Weinberg (1983) & Ruley (1983) extended these
findings to cellular oncogenes activated in human cancer cells, by
demonstrating that transformation of primary rat embryo cells by
a mutated *ras* gene requires the simultaneous transfer of another
oncogene, which can be either polyoma virus *plt*, or adenovirus
E1A, or the rearranged forms of the *myc* gene present in the genome
of MC29 virus (v-*myc*) or cloned from tumor cells (c-*myc*). As
suggested by these results, the rearranged *myc* oncogenes could be
shown to confer on cells in culture, as efficiently as polyoma
plt, serum-independent growth properties and immortalization
(Mougneau *et al.*, 1984).

Taken together, these results indicate that two important stages
can be defined in cell transformation, which are respectively dependent of the expression of viral or activated cellular oncogenes
of two classes: the "immortalizing genes" (polyoma *plt*, adenovirus
E1A, viral and rearranged cellular *myc* genes), the "terminal transformation genes" (polyoma *pmt*, adenovirus E1B, viral and mutated
cellular *ras* genes).

It was of interest to notice that all the genes of the first
class encode nuclear proteins strongly associated with the chromatin of the host cell. These genes are thought to change the
regulation of the expression of important cellular genes, whose

identification is now under way in several laboratories. By contrast, most of the genes of the second group encode membrane-associated proteins, possibly altering the hormonal controls of
the cellular growth (growth factors analogs, proteins associated
with growth factor receptors.

REFERENCES

1) BENJAMIN, T.L. (1970). Host range mutants of polyoma virus.
   Proc. Natl. Acad. Sci. USA, 67, 394.
2) CUZIN, F. (1984). The polyoma virus oncogenes: coordinated
   functions of three distinct proteins in the transformation of
   rodent cells in culture. Biochim. Biophys. Acta, 781, 193.
3) FLUCK, M.M., STANELONI, R.J., & BENJAMIN, T.L. (1977). Hrt
   and ts-a: two early gene functions of polyoma virus. Virology,
   77, 610.
4) FRIED, M. (1965). Isolation of temperature-sensitive mutants
   of polyoma virus. Virology, 25, 669.
5) HOUWELING, A., VAN DEN ELSEN, P.J., & VAN DER EB, A.J. (1980).
   Partial transformation of primary rat cells by the leftmost
   4.5% fragment of adenovirus 5 DNA. Virology, 105, 537.
6) ITO, Y., SPURR, N., & DULBECCO, R. (1977). Characterization
   of polyoma virus T antigen. Proc. Natl. Acad. Sci. USA, 74,
   1259.
7) ITO, Y. (1980). Organization and expression of the genome of
   polyoma virus. In: Viral Oncology (ed. G. Klein), p. 447,
   Raven Press, NY.
8) LAND, H., PARADA, L.F., & WEINBERG, R.A. (1983). Tumorigenic
   conversion of primary embryo fibroblasts requires at least two
   cooperating oncogenes. Nature, 304, 596.
9) MOUGNEAU, E., LEMIEUX, L., RASSOULZADEGAN, M., & CUZIN, F.
   (1984). Biological activities of v-myc and rearranged c-myc
   oncogenes in rat fibroblast cells in culture. Proc. Natl. Acad.
   Sci. USA, in press.

10) RASSOULZADEGAN, M., MOUGNEAU, E., PERBAL, B., GAUDRAY, P., BIRG, F., & CUZIN, F. (1980). Host-virus interactions critical for cellular transformation by polyoma virus and simian virus 40. Cold Spring Harbor Symp. Quant. Biol., 44, 333.

11) RASSOULZADEGAN, M., GAUDRAY, P., CANNING, M., TREJO-AVILA, L., & CUZIN, F. (1981). Two polyoma gene functions involved in the expression of the transformed phenotype in FR3T3 rat cells. I. Localization of a transformation maintenance function in the proximal half of the large-T coding region. Virology, 114, 489.

12) RASSOULZADEGAN, M., COWIE, A., CARR, A., GLAICHENHAUS, N., KAMEN, R., & CUZIN, F. (1982). The roles of individual polyoma virus early proteins in oncogenic transformation. Nature, 300, 713.

13) RASSOULZADEGAN, M., NAGHASHFAR, Z., COWIE, A., CARR, A., GRISONI, M., KAMEN, R., & CUZIN, F. (1983). Expression of the large T protein of polyoma virus promotes the establishment in culture of "normal" rodent fibroblast cells. Proc. Natl. Acad. Sci. USA, 80, 4354.

14) RULEY, H.E. (1983). Adenovirus early region 1A enables viral and cellular transforming genes to transform primary cells in culture. Nature, 304, 602.

15) SCHAFFHAUSEN, B.S., SILVER, J.E., & BENJAMIN, T.L. (1978). Tumor antigen(s) in cells productively infected by wild-type polyoma virus and mutant NG-18. Proc. Natl. Acad. Sci. USA, 75, 79.

16) SEIF, R. & CUZIN, F. (1977). Temperature-sensitive growth regulation in one type of transformed rat cells induced by the tsa mutant of polyoma virus. J. Virol., 24, 721.

17) STEWART, S.E. (1960). The polyoma virus. Adv. Virus Res., 7, 61.

18) TODARO, G.J. & GREEN, H. (1963). Quantitative studies on the growth of mouse embryo cells in culture and their development into established lines. J. Cell Biol., 17, 299.

19) TREISMAN, R., NOVAK, U., FAVALORO, J., & KAMEN, R. (1981). Transformation of rat cells by an altered polyoma virus genome expressing only the middle-T protein. Nature, 292, 595.

20) TYNDALL, C., LA MANTIA, G., THACKER, C.M., FAVALORO, J., & KAMEN, R. (1981). A region of the polyoma virus genome between the replication origin and late protein coding sequences is required in cis for both early gene expression and viral DNA replication. Nucl. Acids Res., 9, 6231.

21) VAN DEN ELSEN, P.J., DE PATER, S., HOUWELING, A., VAN DER VEER, J., & VAN DER EB, A. (1982). The relationship between region E1a and E1b of human adenoviruses in cell transformation. Gene, 18, 175.

22) VOGT, M. & DULBECCO, R. (1962). Studies on cells rendered neoplastic by polyoma virus. Virology, 16, 41.

23) VOGT, M. & DULBECCO, R. (1963). Steps in the neoplastic transformation of hamster embryo cells by polyoma virus. Proc. Natl. Acad. Sci. USA, 49, 171.

PAPOVA VIRUSES AND CANCER GENES

C. Streuli* & Beverly E. Griffin#

*Imperial Cancer Research Fund, Lincoln's Inn Fields
London WC2, England
#Department of Virology, Royal Postgraduate Medical
School, Hammersmith Hospital, London W12, England

INTRODUCTION

The group of small, double-stranded DNA tumour viruses, classified
as papova viruses, have received considerable attention following
the discovery of the first member of this species, papilloma virus,
in 1933 by Shope in rabbits. Since Shope's discovery, *papilloma*
viruses have been recovered from the warts of many other animals,
including humans (for review, see reference 1). The other members
of the species which have provided the name for this class of
viruses, *po*lyoma, a mouse virus, and the *va*cuolating virus of rhesus
monkeys, SV40, were not isolated until 1953 and 1960, respectively
(for review, see references 2-4). Since then a number of human
papova viruses have been discovered (for review, see reference 5)
and most recently, a papovavirus of hamsters (6) was isolated. Thus
this class of viruses, all capable of producing cellular transformation in appropriate tissue culture systems, or tumours in suitable
animal models, is ubiquitous throughout the animal kingdom. The
association with malignancies has provided the stimulus for many
different studies. The very small sizes of the viruses (between
about 5,000 and 10,000 base pairs) offers the possibility that the

135

virus-host cell interactions that ultimately result in production
of a cancer-like cell can eventually be elucidated. Although this
provides an important rationale for many of the investigations
to date, the further discovery that the viral DNAs exist in the
nuclei of their hosts as 'mini-chromosomes' (7), dependent upon,
and mimicking, host cell machinery, has focussed attention on their
use as models for studying regulatory processes active in normal
cells. In this regard, 'enhancer sequences' which play an important
role in the control of transcription, 'TATA and CAT boxes' which
act as promoter signals and the splicing phenomena that produce
mRNAs have been most thoroughly characterised in viral genomes. In
the case of the former, it may be relevant that very little sequence
homology has been found among enhancers of different viruses (8).

Our present knowledge would suggest that whereas most members of the
papova viruses lend themselves to ready and useful comparisons, the
papillomas stand apart. Not only are papilloma viruses nearly twice
as large as the other members of the species, they show little
homology with them at the DNA level (1). Moreover, unlike the latter
(polyoma virus, SV40, BKV, etc.) they exist in their host cells as
episomal (autonomously replicating) moieties exerting control of
gene expression without integrating into their chromosomes. Indeed,
a sequence that appears actively to prevent integration has been
identified (9). No lytic system has yet been found for these viruses.
The recent surge of interest in papilloma viruses has been a consequence of the discovery of an association between human papilloma
viruses and cervical cancer (10).

As mentioned, with the exception of papillomas, the other papova
viruses are generally similar in size, sequence and coding capacity.
All appear to integrate into their host cell chromosomes during the
process(es) that lead to cellular transformation. It is important
to note that in their natural hosts, these viruses in general invoke

a lytic response, and transformation is reserved either to heterologous host species or to interactions between mutant viruses and
natural hosts. To date, the best studied of the viruses are polyoma
and SV40. The following section therefore focusses briefly on the
relationship between these two viruses. Since lytic systems are
available for both, they are relatively easily studied and in terms
of cellular transformation, they are better understood than the
papilloma viruses.

COMPARISONS OF POLYOMA VIRUS AND SV40

Early experimental comparisons of polyoma virus and SV40 involved
heteroduplex analysis by electron microscopy (11), DNA-DNA hybridisation studies (12), and determination of immunological cross-
reactivities between viral gene products (13,14). The data were
entirely consistent. That is, the strongest degree of homology
existed between regions coding for portions of the viral capsid
proteins, with a second area of homology in a region now known to
encode the C-termini of the large T-antigens (LT) of both species,
as illustrated (Figure 1). The determination of the complete DNA
sequences of one strain of SV40 in 1978 (15,16) and two strains of
polyoma virus in 1980 (17,18) allowed very precise comparisons
between the viral genomes and their predicted proteins to be made
(Figure 2). Thus, it is obvious that in most of the genomic regions
specifying known viral proteins, considerable homology between the
two viruses both at the DNA and protein levels exist. There are two
major obvious regions of inhomology. One lies in the portion of DNA
that encodes the mid- and C-terminal regions of the middle T-antigen unique to polyoma virus (see below), and the other in the noncoding 'control' regions of the two viruses which contain sequences
that specify origins of replication and transcriptional regulatory
functions, including enhancers (reviewed, reference 19). (It would
not be surprising if the latter prove in each case to resemble
sequences found in the hosts of the individual viruses, that is,

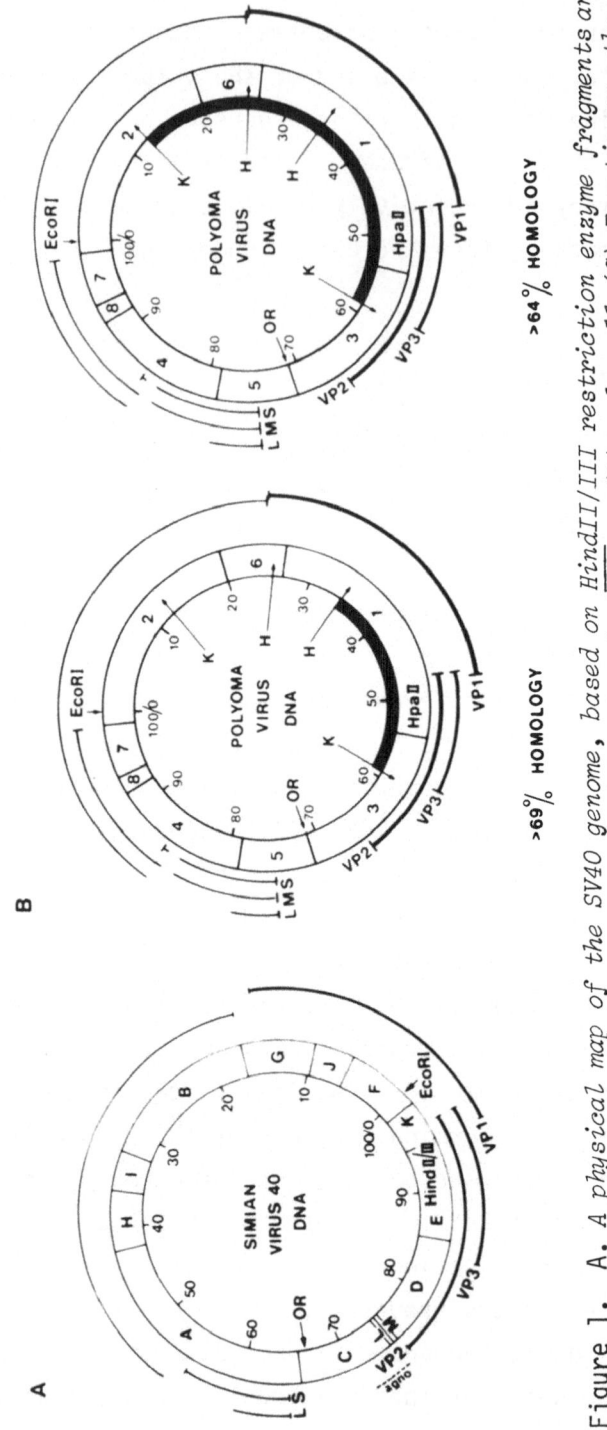

Figure 1. A. A physical map of the SV40 genome, based on HindII/III restriction enzyme fragments and divided into 100 units, showing the locations of the viral large (L) and small (S) T-antigens, the capsid proteins (VP1, VP2, VP3) and the origin of replication (OR). Data taken from the DNA sequence (15,16). Similarly, B. A physical map of the polyoma virus genome, based on HpaII restriction enzyme fragments, showing the locations of the viral large (L), middle (M), and small (S) T-antigens, the capsid proteins and the origin of replication. The heavy bands show homologies between the DNAs of the two viruses detected by DNA hybridisation under stringent conditions which allowed only detection of >69% homology, and less stringent (>64% homology) conditions (12). The two viral genomes are aligned in such a way as to allow direct comparison.

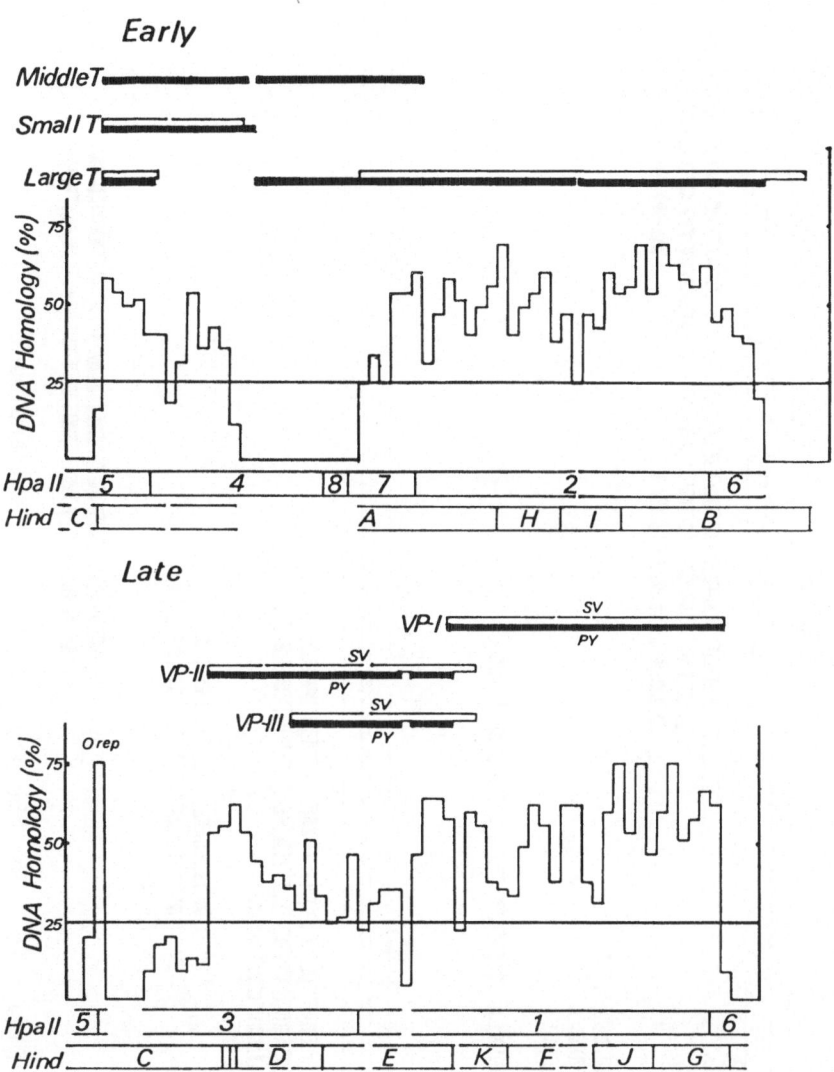

Figure 2. *Schematic diagrams showing areas of homology and inhomo-*
logy between the genomes of SV40 and polyoma virus in terms of, A,
their DNA sequences, and, B, their predicted protein sequences.
In the latter, a dark vertical line is used to indicate conserva-
tion of an amino acid sequence between the two species.
(Figure taken from Soeda *et al*. (18).)

See overleaf for Figure 2B.

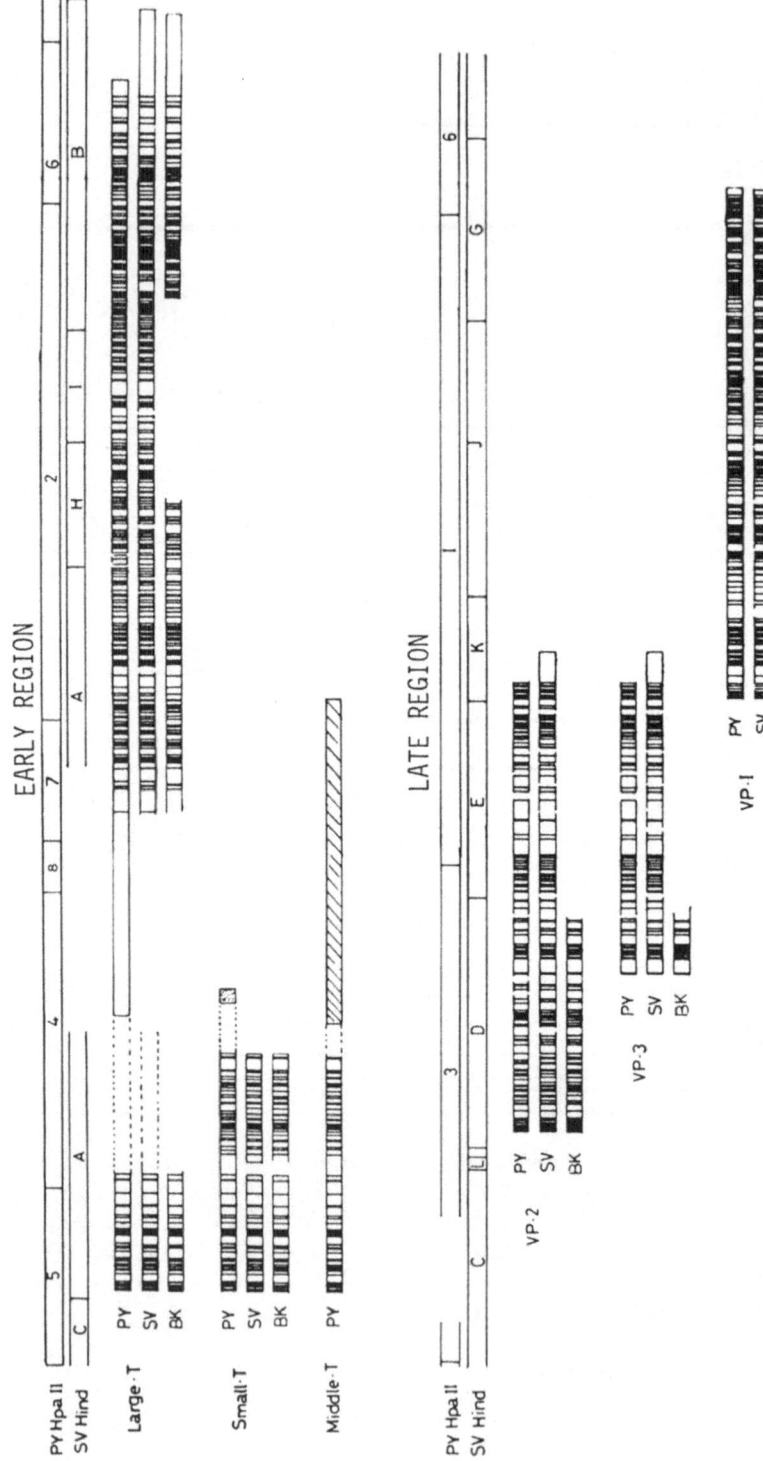

Figure 2B.   See the previous page for explanation.

mouse DNA in the case of polyoma and monkey DNA in the case of SV40,
since the viruses in part regulate and in part depend for their
existence on the expression of host cell functions). When the
sequence of a human papova virus, BKV, was determined it was found
to resemble SV40 more than polyoma virus (20), both in terms of its
genetic organisation and DNA/protein composition.

So far, the homologies between polyoma virus and SV40 have been
discussed chiefly in terms of information obtained from studies of
their physical properties. Biological studies have also shown the
viruses to have many similar properties. In both cases, the cellular transformation appears to be a consequence of the expression of
viral gene products. The genomes of both can be divided into 'early'
and 'late' transcription regions, that is, regions that are expressed before DNA replication can be detected, and regions expressed post-
replication, respectively (for review, see reference 21). The former
are known to specify proteins that are designated as tumour-(or T-)
antigens, because they are generally found in lesions produced in
animals in response to viral infections and can be immunoprecipitated by anti-tumour antisera; the lattter specify the three characterised viral capsid proteins (reviewed in reference 22). The
surprise in this field was the discovery of a protein unique to
polyoma virus (23); this was designated *middle* T-antigen because of
its size (55K), which lies between that of the two previously identified T-antigens, the *large* (90-100K) and *small* (22-23K) T-antigens,
specified by both polyoma virus and SV40. In spite of very considerable searches, no SV40 (or BKV) middle T-antigen equivalent has
ever been discovered. DNA sequence data (15,16,20) suggest that it
may not exist.

CELLULAR TRANSFORMATION BY POLYOMA VIRUS

Many different types of studies led to the notion that *in vitro* the
55K middle T-antigen of polyoma virus was primarily responsible for

inducing a fully-transformed state in virally-infected cells. Much
of the early evidence was circumstantial, but when combined was
nonetheless highly compelling. This is summarised as follows:

a) A series of viral mutants, designated host-range non-transforming or *hrt*, have been isolated and characterised that lack the
ability to transform cells in culture (24). These variants, many of
which have large deletions and are frame-shift mutants, do not express a normal middle (or small) T-antigen (25); all these mutants
make a large T-antigen that should be functional. A single member
of this class, NG-59, apparently expresses a full-sized middle T-
antigen with a two-amino acid alteration; it is also non-transforming
(26).

b) Another series of viral mutants, designated *mlt*, have modifications in the overlapping regions that code for portions of both
middle and large T-antigens; these do not affect expression of small
T-antigen (reviewed in 19). Many of these mutants are transformation deficient (27). Interestingly one, dl 8, transforms cells in
culture and produces tumours in whole animals more efficiently than
wild type virus (28).

c) A region of the DNA that encompasses only a third of the viral
genome (Figure 3), which retains the capacity for middle (and small)
T-antigen expression, is fully competent to transform cells (29).
Any attempts to diminish the size of this fragment in such a way as
to impinge upon the expression of middle T-antigen was found to
have a dramatic (negative) effect upon transformation (30).

Evidence against a primary role for large T-antigen in transformation, at least in some instances, comes from two different types
of experiments:

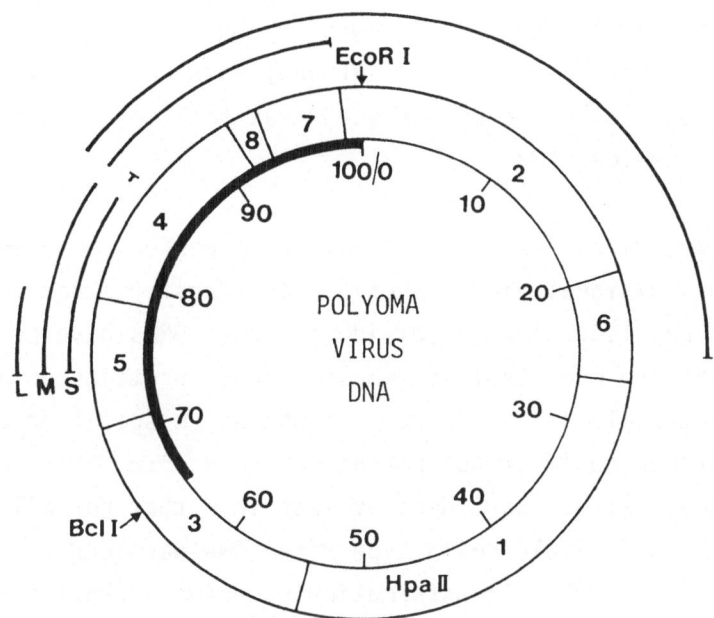

<u>Figure 3</u>. *Transformation by a sub-genomic fragment of polyoma virus DNA*. A segment of the polyoma genome encompassed within the smaller of the two fragments generated by cleavage with restriction enzymes *BCl*I and *Eco*RI, as illustrated (heavy area) was found to be fully competent to transform cells in culture and produce tumours in whole animals (Novak *et al*., reference 29). Alterations to the DNA in the region around 98 map units (C-terminus of middle T-antigen, abolished transformation (30)).

a) The *tsa* mutants have temperature sensitive large T-antigens (22) and, as virions, can only induce cellular transformation at the permissive (32°), not at the non-permissive (39°), temperature. Nonetheless, it has been shown that DNA isolated from such mutants, when transfected into cells at either temperature, produce dense foci (31). Thus, transformation by these mutants may be only indirectly related to the lesions in the viral antigen.

b) Many cell lines derived either from densely-growing foci *in*

vitro or from virally-induced tumours in whole animals fail to
express full-sized viral large T-antigen; no stable transformed
cell lines to date have been analysed that fail to express middle
(or small) T-antigen (32).

More recently, direct experiments have been performed to test the
rôle of the individual viral functions in transformation. Using
recombinant DNA technology, individual viral cDNAs have been
produced that are competent to express single proteins only (33).
With such material, it was found that DNA which specified only
middle T-antigen could transform rat cells *in vitro*, although with
an efficiency that was considerably less than that normally observed when, for example, wild type viral DNA was used. It was
subsequently shown that transformation of primary (rat) cells requires the expression of some other viral function, which was defined in one case as the N-terminus of the large T-antigen (34),
and in another, as small T-antigen (35). In newborn hamsters, expression of middle T-antigen alone has been found to be competent
for tumour induction (35).

In many of the above experiments there is an inherent problem in
that the host cell being acted upon by the virus is often not well-
defined, and comparisons and deductions regarding viral function are
sometimes made among experiments without due consideration for the
nature of the host cell and its environment. This very basic problem is now being recognised, however, and given some attention
since the virus itself has been well-characterised to the extent
that the next obvious stage of study is its interaction with its
host cell. Nonetheless, experiments to date overall emphasise the
key role of polyoma virus middle T-antigen in cell transformation,
but leave largely unresolved the secondary but possibly highly
relevant roles played by the other antigens in this cellular event.
(For a recent review, see references 19,36,37).

In the case of SV40, where expression of the large T-antigen is of
undoubted importance in achieving cellular transformation, where
small T-antigen may be relevant to achieving full-transformation,
and where there is no middle T-antigen, host cell factors may be of
even greater relevance than in the case with polyoma virus (37,38).

It is becoming apparent that different forms of the viral T-antigens
exist and may be of biological significance. For example, two different phosphorylated forms of polyoma virus large T-antigen have
been identified, and in many cell lines truncated forms of the protein also exist (for review, see references 19,36,37). The relevance of these (presumably) post-translational modifications remains to be determined. It is noteworthy that in the case of SV40,
both nuclear and membrane-associated forms of the large T-antigen
exist (39). Similarly, small T-antigen has been identified both in
the cytoplasm and in the nucleus (40). In the case of polyoma virus,
many truncated species related to the C-terminus of middle T-antigen have been identified, especially in lytically infected cells,
and a prominent 60K related species can be immunoprecipitated from
both lytically-infected and tumour-derived cells (41, our unpublished results) (see Figure 4). Since immunoelectron microscopy has
shown middle T-antigen(s) to associate with all cellular membranes
(19, our preliminary data), the possibility of different forms of
the antigen exerting specific effects at numerous cell sites cannot
be ignored; similar arguments could be made for SV40. Dilworth,
using monoclonal antibodies, showed that the protein kinase activity
associated with middle T-antigen (41) resided in an immunological
sub-population of this antigen, providing another argument in favour
of more than one form of the antigen, and possibly different functions.

It may in time prove that these very small viruses, which are already known to use differential splicing as a means of overcoming

their limited coding capacity, are found to have multi-functional
proteins created by protein processing or modification.

NEW EXPERIMENTS

We have carried out several experiments recently that are aimed at
elucidating functions for the individual T-antigens of polyoma
virus and investigating the origin of those sequences unique to
this virus, that is, most of the region that specifies the middle
and C-terminus of middle T-antigen (see Figure 2B). Our studies
have involved cells of murine origin only, in an attempt to mimic
in as far as possible normal viral-cell interactions by focussing
on the natural host of the virus. Two of these will be briefly discussed:

a) DNA complements of messenger RNAs (that is, cDNA) that specify
a single early function only (33) as well as late, capsid proteins
were individually transfected into primary mouse embryo or mouse
kidney cells, and an assay for virus production was carried out. In
both cell populations, virus was produced in the situation which
allowed, among the early gene products, only for the expression of
large T-antigen, and the capsid proteins; virions were not produced
in similar experiments where small and middle T-antigens would be
expressed, together with capsids. That is, a polyoma-related virus
can be produced that expresses only four of the six known viral
proteins, namely large T-antigen, VP1, VP2, and VP3. This virus is
non-transforming. Moreover, it is not as infectious as wild type
virus in that it produced small plaques only whereas, under similar
conditions, large plaques developed following wild type viral infection. Cotransfections with mixtures of cDNAs did not restore
the large plaque phenotype. To distinguish among the various factors that could account for this failure will require further genetic manipulations which have not yet been made. The experiment *per*

se is important in that it allows for the possibility that in ancestral time a lytic polyoma-related virus may have existed which
acquired a sequence from its host cell to produce the present wild
type transforming virus by a step-wise progression, somewhat analogous to that which is now well-known in the retrovirus field,
where proto-oncogenes have been acquired from the host by otherwise
nontransforming viruses, to generate transforming species. The difference is that in the latter case, during the acquisition of the
'oncogene', sequences have also been lost that result in defective
transforming viruses which require helper viruses for propagation.
From the above data, this would obviously not be the case with polyoma virus. Precedence for the acquisition of host sequences by polyoma virus during its growth cycle in mouse cells may be found in
the work of Bourgaux and collaborators (43, and references therein)
and Ding *et al.* (44).

b)   The second experiment had two goals in mind: One was to ascer-
tain whether an underlying 'second mode' of transformation exists
in polyoma virus, dependent upon the expression of large T-antigen.
That is, if in the absence of middle T, by analogy with SV40 (see
Figure 1), polyoma virus could transform cells in its natural host
making use of functions expressed by its large T-antigen. If not,
to ask the question whether sequences could be acquired from the
host to rescue a defect in transformation. The experiment chosen
initially, for which data are now available (unpublished), was to
inject nude (nu/nu) mice with the *hrt* mutant, NG18 (which should not
specify either middle or small T-antigen, but should express a wild
type large T-antigen (25)) that is non-transforming *in vitro*.
Whereas under similar conditions, mice injected with wild type virus
should produce tumours within a matter of weeks, the first tumours
appeared in mice injected with NG-18 about eight months post-infect-
ion. Tumours continued to appear in animals up to 15 months following the initial viral innoculation, after which time the experiment
was terminated. In two different experiments, lesions occurred in

8/9 and 6/10 animals. Attempts to establish cell lines in culture from tumour materials met with partial success, and five lines were eventually established. It is noteworthy that although the mutant NG18 itself appears to be defective also in terms of lytic growth, particularly in mouse fibroblast cells (24), in our experiments most cell populations derived from tumours did produce virus, which may account for the failure of many cells to become established *in vitro*. Analysis of the total chromosomal DNA in established lines, following cleavage with restriction endonucleases, showed several of them to contain many copies of a defective viral species unlike any studied to date (3). Another line showed a *Hpa*II restriction pattern identical to that observed by Feunteun *et al.* (45) in experiments aimed at 'marker rescue' of the wild type phenotype from *hrt* mutants. A third line, analysed in more detail, contained multicopies of episomal mutant NG18 genomes which obscured analysis of integrated viral DNA. Continued passage in culture resulted in the loss of episomal DNA, and in preliminary experiments revealed an underlying chromosomal DNA pattern that could be shown by 'Southernblot' analysis to resemble completely neither a wild type, revertant, nor NG-18 mutant DNA. All lines established from tumours produced a 54K (as opposed to wild type 55K) middle T-antigen, and small T-antigen, as well as reasonable amounts of a 60K middle T-antigen-related species (Figure 4). None of the lines produced a full-sized large T-antigen. Cells re-injected into nude mice were found to be highly tumorigenic, producing lesions in a matter of days, mainly at the site of injection.

The simplest conclusion to be drawn from these experiments is that tumour formation probably did not occur as a consequence of large T-antigen expression. Rather, the virus appears to have 'rescued' a transforming function from its host. Whether this has involved the acquisition of specific murine ('oncogene') sequences or of arbitrary mouse sequences which merely restore the middle T-antigen

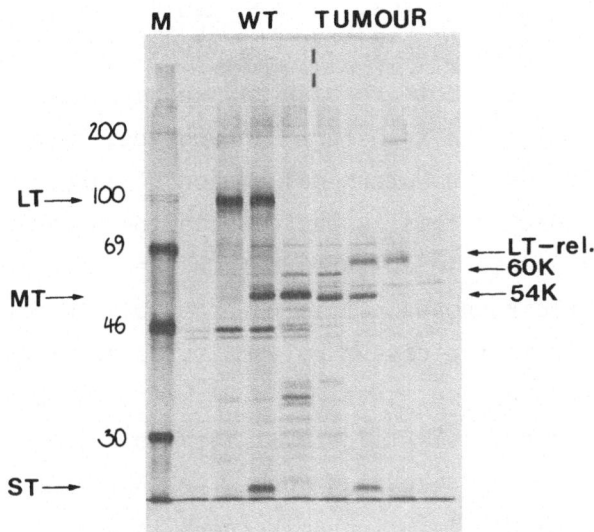

Figure 4. *Immunoprecipitation and separation (SDS-PAGE) of* ^{35}S-*labelled proteins from mouse cells lytically infected with wild type virus (WT) and from mouse tumour (TUMOUR) cells.* Immunoprecipitations were carried out using monoclonal antibodies specific for large T-antigen (LT4), for the common region of the three antigens (C4, see Figure 3) and for middle T-antigen (MT16), as described (46). The locations of the three wild type (WT) antigens (LT, MT, ST) are designated by arrows (at left) and of the related proteins prominent in tumours by arrows (at right). The 'family' of smaller polypeptides (4) related to middle T-antigen may be observed in the track of lytically infected cells immunoprecipitated with MT16. Protein size markers (M) are included.

reading frame, or whether the crucial event for tumour formation has been the consequence of viral activation of a host gene such as, for example, the 60K species, remains to be seen. The data obtained in this study support the notion, which still remains to be proved, that an essentially lytic form of polyoma virus may in past ages have been converted into a transforming virus by the acquisition of genetic information from its (mouse) host.

CONCLUSIONS

Each of the two papova viruses studied to date in detail encode a transforming, or cancer, gene. In the case of SV40, this appears to be its large T-antigen, and in polyoma virus, the middle T-antigen. In heterologous hosts, both viruses may require other functions for full-expression of a transformed phenotype, such as small T-antigen for SV40, or the N-terminal region of large T-antigen for polyoma virus. Whether this is the case in the natural host is difficult to prove, since in culture these viruses are fully-competent to produce a lytic response. It is now possible to approach this question directly in the case of polyoma virus by transfecting genetically engineered genomes into whole animals, but only preliminary results from such experiments are yet available (35, our unpublished data). Our results, discussed above, further emphasise the need to search for a middle T-antigen-like gene in normal mouse cells.

REFERENCES

1) ZUR HAUSEN, H. (1981). Papilloma viruses. In: "DNA Tumor Viruses" J. Tooze, ed., Second edition, Part 2, revised. Cold Spring Harbor, New York, p. 371.

2) PONTÉN, J. (1971). Spontaneous and virus induced transformation in cell culture. In: "Virology Monographs", S. Gard, C. Hallauer & K.F. Meyer, eds.

3) GRIFFIN, B.E. (1981). Structure and genomic organization of SV40 and polyoma virus. In: "DNA Tumor Viruses", J. Tooze, ed., Second edition, Part 2, revised, Cold Spring Harbor, New York, p. 205.

5) PADGETT, B. (1981). Human papova viruses. In: "DNA Tumor Viruses", J. Tooze, ed., Second edition, Part 2, revised. Cold Spring Harbor, New York, p. 339.

6) SCHERNECK, S., BÖTTGER, M. & FEUNTEUN, J. (1979). Studies on

the DNA of an oncogenic papova virus of Syrian Hamster. Virol.
96, 100.

7) GRIFFITH, J. (1975). Chromatin structure: Deduced from a mini-
chromosome. Science, 187, 1202; CREMISI, C., PIGNATI, P.F.,
CROISSANT, O. & YANIV, M. (1976). Chromatin-like structures in
polyoma virus and simian virus 40 lytic cycle. J. Virol., 17,
204.

8) GRUSS, P., DHAR, R. & KHOURY, G. (1981). Simian virus 40 tandem repeated sequences as an element of the early promoter,
Proc. Natl. Acad. Sci. USA, 78, 943;
BENOIST, C. & CHAMBON, P. (1981). In vivo sequence requirements of the SV40 early promoter region, Nature, 290, 304.
DE VILLIERS, J. & SCHAFFNER, W. (1981). A small segment of
polyoma virus DNA enhances the expression of a cloned β-globin
gene over a distance of 1400 base pairs, Nucl. Acids Res., 9,
6251.
SPANDIDOS, D.A. & WILKIE, N.M. (1983). Host-specificities of
papilloma virus, moloney sarcoma virus and simian virus 40 enhancer sequences, EMBO J., 2, 1193.

9) LUSKY, M. & BOTCHAN, M.R. (1984). Characterization of the
bovine papilloma virus plasmid maintenance sequences. Cell, 36,
391.

10) DÜRST, M., GISSMANN, L., IKENBERG, H. & ZUR HAUSEN, H. (1983).
A papilloma DNA from a cervical carcinoma and its prevalence
in cancer biopsy samples from different geographical regions.
Proc. Natl. Acad. Sci. USA, 80, 3812.

11) FERGUSON, J. & DAVIS, R.W. (1975). An electron microscopic
method for studying and mapping the region of weak sequence
homology between simian virus 40 and polyoma DNAs. J. Mol.
Biol., 94, 135.

12) HOWLEY, P.M., ISRAEL, M.A., LAW, M.-F. & MARTIN, M.A. (1979).
A rapid method for detecting and mapping homology betweeen
heterologous DNAs. J. Biol. Chem., 254, 4876.

13) SHAH, K.V., OZER, H.L., GHAZEY, H.N. & KELLY, T.J. (Jr)(1977). Common structural antigen of papova viruses of the simian virus 40 - polyoma subgroup. J. Virol., 21, 179.

14) McCORMICK, F., LANE, D.P. & DILWORTH, S.M. (1982). Immunological cross-reaction between large T-antigens of SV40 and polyoma virus. Virol., 116, 382.

15) FIERS, W., CONTRERAS, R., HAEGEMAN, G., ROGIERS, R., VAN DE VOORDE, E., VAN HERREWEGHE, J., VOLCKAERT, G. & YSEBAERT, M. (1978). Complete nucleotide sequence of SV40 DNA. Nature, 273, 113.

16) REDDY, V.B., THIMMAPPAYA, B., DHAR, T., SUBRAMANIAN, K.N., ZAIN, B.S., PAN, J., GHOSH, P.K., CELMA, M.L. & WEISSMAN, S.M. (1978). The genome of simian virus 40. Science, 200, 494.

17) DEININGER, P. ESTY, A., LaPORTE, P., HSU, H. & FRIEDMAN, T. (1981). The nucleotide sequence and restriction enzyme sites of the polyoma genome. Nucl. Acids Res., 8, 855.

18) SOEDA, E., ARRAND, J.R., SMOLAR, N., WALSH, J. & GRIFFIN, B.E. (1980). Coding potential and regulatory signals of the polyoma virus genome. Nature, 283, 445.

19) GRIFFIN, B.E. & DILWORTH, S.M. (1983). Polyoma virus: An overview of its unique properties. In: "Advances in Cancer Research" G. Klein & S. Weinhouse, eds., Academic Press, Inc., New York, p. 183.

20) YANG, R.C.A. & WU, R. (1979). BK virus DNA: complete nucleotide sequence of a human tumor virus. Science, 206, 456.

21) ACHESON, N.H. (1981). Lytic cycle of SV40 and polyoma virus. In: "DNA Tumor Viruses", J. Tooze, ed., Second edition, Part 2, revised, Cold Spring Harbor, New York, p. 125.

22) TEGTMEYER, P. (1981). Genetics of SV40 and polyoma virus. In: "DNA Tumor Viruses", J. Tooze, ed., Second edition, Part 2, revised, Cold Spring Harbor, New York, p. 297.

23) ITO, Y., BROCKLEHURST, J.T. & DULBECCO, R. (1977). Virus-specific proteins in the plasma membrane of cells lytically

infected or transformed by polyoma virus. Proc. Natl. Acad.
Sci. USA, 74, 4666.

24) BENJAMIN, T.L. (1970). Host range mutants of polyoma virus.
Proc. Natl. Acad. Sci. USA, 67, 394; STANELONI, R.J., FLUCK,
M.M. & BENJAMIN, T.L. (1977). Host range selection of trans-
formation-defective hr-t mutants of polyoma virus. Virol., 77,
598.

25) SOEDA, E. & GRIFFIN, B.E. (1978). Sequences from the genome of
a non-transforming mutant of polyoma virus. Nature, 276, 294;
SILVER, J., SCHAFFHAUSEN, B. & BENJAMIN, T. (1978). Tumor
antigens induced by non-transforming mutants of polyoma virus.
Cell, 15, 485.

26) CARMICHAEL, G.G. & BENJAMIN, T.L. (1980). Identification of
DNA sequence changes leading to loss of transforming ability
in polyoma virus. J. Biol. Chem., 255, 230.

27) DING, D., DILWORTH, S.M. & GRIFFIN, B.E. (1982). mlt mutants
of polyoma virus. J. Virol., 44, 1080.

28) GRIFFIN, B.E. & MADDOCK, C. (1979). New classes of viable
deletion mutants in the early region of polyoma virus. J.
Virol., 31, 645; GRIFFIN, B.E., ITO, Y., NOVAK, U., SPURR, N.,
DILWORTH, S.M., SMOLAR, N., POLLACK, R., SMITH, K. & RIFKIN,
D.B. (1980). Early mutants of polyoma virus (dl 8 and dl 23)
with altered transformation properties: Is polyoma virus middle
T-antigen a transforming gene product? Cold Spring Harbor Symp.
Quant. Biol., 44, 271.

29) CHOUDHURY, K., LIGHT, S.E., GARON, C.F., ITO, Y. & ISRAEL, M.
A. (1980). A cloned polyoma DNA fragment representing the 5'
half of the early gene is oncogenic. J. Virol., 36, 566;
HASSELL, J.A., TOPP, W.E., RIFKIN, D.B. & MOREAU, P.E. (1980).
Transformation of rat embryo fibroblasts by cloned polyoma
virus DNA fragments containing only part of the early region.
Proc. Natl. Acad. Sci. USA, 77, 3978; NOVAK, U., DILWORTH,
S.M. & GRIFFIN, B.E. (1980). Coding capacity of a 35% fragment

of the polyoma virus genome is sufficient to initiate and maintain cellular transformation. *ibid.*, 77, 3278.

30) NOVAK, U. & GRIFFIN, B.E. (1981). Requirement for the C-terminal region of middle T-antigen in cellular transformation by
polyoma virus. Nucl. Acids Res., 9, 2055; CARMICHAEL, G.G.,
SCHAFFHAUSEN, B.F., DORSKY, K.I., OLIVER, D.B. & BENJAMIN, T.
L. (1982). Carboxy terminus of polyoma middle-sized tumor
antigen is required for attachment to membranes, associated
protein kinase activities and cell transformation. Proc. Natl.
Acad. Sci., 79, 3579.

31) NOVAK, U. & GRIFFIN, B.E. (1981). Cellular transformation by
polyoma virus. In: "Int. Cell Biology 1980-1981". H.G.
Schweiger, ed., Springer-Verlag, Berlin, p. 448.

32) ITO, Y. & SPURR, N. (1980). Polyoma virus T antigens expressed
in transformed cells: Significance of middle T antigen in
transformation. Cold Spring Harbor Symp. Quant. Biol., 44, 149.

33) TREISMAN, R., NOVAK, U., FAVALORO, J. & KAMEN, R. (1981).
Transformation of rat cells by an altered polyoma virus genome
expressing only the middle-T protein. Nature, 292, 595; ZHU,
Z., VELDMAN, G.M., COWIE, A., CARR, A., SCHAFFHAUSEN, B. &
KAMEN, R. (1984). Construction and functional characterization
of polyoma virus genomes that separately encode the three early
antigens. J. Virol., 51, 170.

34) RASSOULZADEGAN, M., NAGHASHFAR, Z., COWIE, A., CARR, A.,
GRISONI, M., KAMEN, R. & CUZIN, F. (1983). Expression of the
large T protein of polyoma virus promotes the establishment
in culture of 'normal' rodent fibroblast cell lines. Proc.
Natl. Acad. Sci. USA, 80, 4354.

35) ASSELIN, C., GÉLINAS, C., BRANTON, P.E. & BASTIN, M. (1984).
Polyoma middle T antigen requires cooperation from another
gene to express the malignant phenotype *in vivo*. Mol. and
Cell. Biol., 4, 755.

36) CUZIN, F. (1984). The polyoma virus oncogenes. Co-ordinated
    functions of three distinct proteins in the transformation
    of rodent cells in culture. Biochim. Biophys. Acta, 781, 193.

37) SCHAFFHAUSEN, B. (1982). Transforming genes and gene products
    of polyoma and SV40. In: "CRC Crit. Rev. Biochem.", 13, 215.

38) CLAYTON, C.E., MURPHY, D., LOVETT, M. & RIGBY, P.W. (1982).
    A fragment of the SV40 large T-antigen gene transforms. Nature,
    299, 59; COLBY, W.W. & SHENK, T. (1982). Fragments of the
    simian virus 40 transforming gene facilitate transformation
    of rat embryo cells. Proc. Natl. Acad. Sci. USA, 79, 5189;
    GRAESSMANN, M., GRAESSMANN, A. & MUELLER, C. (1980). Monkey
    cells transformed by SV40 DNA fragments: Flat revertants syn-
    thesize large and small T antigens. Cold Spring Harbor Symp.
    Quant. Biol., 44, 605.

39) DEPPERT, W., HANKE, K. & HENNING, R. (1980). Simian virus 40
    T-antigen-related cell surface antigen: serological demonstra-
    tion on simian virus 40-transformed monolayer cells *in situ*.
    J. Virol., 35, 505; SANTOS, M. & BUTEL, J. (1982). Detection
    of a complex of SV40 large tumor antigen and 53K cellular
    protein on the surface of SV40-transformed mouse cells. J.
    Cell. Biochem., 19, 127; STAUFENBIEL, M. & DEPPERT, W. (1983).
    Different structural systems of the nucleus are targets for
    SV40 large T-antigen. Cell, 33, 173.

40) ELLMAN, M., BIKEL, I., FIGGE, J., ROBERTS, T., SCHLOSSMAN, R.
    & LIVINGSTON, M. (1984). Localization of the simian virus 40
    small T antigen in the nucleus and cytoplasm of monkey and
    mouse cells. J. Virol., 50, 623.

41) DILWORTH, S.M. & GRIFFIN, B.E. (1983). The transforming gene
    of polyoma virus. Arch. Geschulstforsch., 53, 187.

42) DILWORTH, S.M. (1983). Protein kinase activities associated
    with distinct antigenic forms of polyoma virus middle T anti-
    gen. EMBO J., 1, 1319.

43) BOURGAUX, P., SYLLA, B.S. & CHARTRAND, P. (1982). Excision of
     polyoma virus DNA from that of a transformed mouse cell: Iden-
     tification of a hybrid molecule with direct and inverted
     repeat sequences at the viral-cellular joints. Virol., 122,
     84.

44) DING, D., JONES, M.D., LEIGH-BROWN, A. & GRIFFIN, B.E. (1982).
     Mutant din-21, a variant of polyoma virus containing a mouse
     DNA sequence in the viral genome. EMBO J., 1, 461.

45) FEUNTEUN, J., SOMPAYRAC, L., FLUCK, M. & BENJAMIN, T. (1976).
     Localization of gene functions in polyoma virus DNA. Proc.
     Natl. Acad. Sci. USA, 73, 4169.

46) DILWORTH, S.M. & GRIFFIN, B.E. (1982). Monoclonal antibodies
     against polyoma virus tumor antigens. Proc. Natl., Acad. Sci.,
     79, 1059.

EPSTEIN-BARR VIRUS AND IMMORTALISATION OF EPITHELIAL CELLS

Beverly E. Griffin
Department of Virology, Royal Postgraduate Medical
School, Hammersmith Hospital, London W12, England

INTRODUCTION

The association of Epstein-Barr virus (EBV) with two human malignancies, the B-cell (or Burkitt's) lymphomas mainly of African
origin, and carcinomas of the nasopharynx (NPC) common to southern
China, as well as its causative role in infectious mononucleosis is
well-documented (1,2). Now being recognised and deserving equal
attention are the EBV-associated lymphoproliferative disorders
arising in immunosuppressed persons, notably among recipients of
bone-marrow, cardiac and renal transplants, many of which prove
fatal (3, and references therein). In spite of the medical importance of this member of the herpes virus family, its precise role in
cellular transformation has proved difficult to elucidate.

Here, we present some experiments carried out in a direct attempt
to localise a transforming gene(s) assuming such exists, within the
viral genome. For technical reasons, epithelial cells have been
used in all our studies to date and it is not yet clear whether our
finding of an immortalising sub-genomic fragment of EBV, suggestive
of an inherent immortalising gene, can be extrapolated to include
B-lymphocytes. There is no doubt (4) that EB virions can immortal-

157

ise B-lymphocytes *in vitro* (but, interestingly, epithelial cells
have not yet been immortalised by virions) although it remains unclear whether the pathway taken for the two cell types is identical.

TRANSFECTION STUDIES ON EPITHELIAL CELLS

In order to perform our studies in such a way as to minimise the
number of experiments involved and, at the same time, reduce the
risk of cleaving within a vital gene and thus destroying it, we
created an EBV cosmid library (5). (See Figure 1).

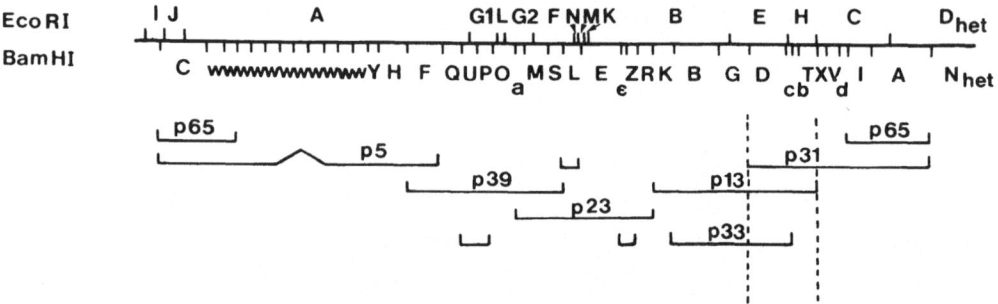

Figure 1. *A recombinant DNA cosmid library derived from partial
cleavage of B95-8 virion DNA with the restriction enzyme, BamHI.*
The fragments that compose the library are illustrated relative to
the *Bam*HI and *Eco*RI standard restriction enzyme maps (19). The
region of overlap between cosmids p13 and p31 is indicated.

DNA derived from EB virions isolated from B95-8 cells (1) was sub-
jected to partial cleavage with the restriction enzyme *Bam*HI and
fragments cloned within the cosmid vector, pHC 79 (6). In this
manner, a recombinant DNA 'library' was obtained consisting of six
essentially overlapping fragments, as shown (Figure 1). These in
turn were used to transfect rodent fibroblasts, and a mixture of
fibroblast and epithelial cells derived from African green monkey
kidneys (AGMK), using the calcium phosphate protocol (7). (This is
a favoured method for introducing DNA into fibroblasts; it is generally lethal for B-lymphocytes but has proved useful for epithel-

ial cells.) In our hands, transfection with either of two overlapping EBV fragments, designated p13 (p33) and p31 into AGMK cells
produced dense cellular growth after several months in culture.
Individual 'foci' were selected for the establishement of cell
lines. Similar growth response was not observed with rodent fibroblasts, nor with AGMK cells transfected with any of the other members of the cosmid library, with EB virions, or with full-length
episomal EBV DNA from a Burkitt's lymphoma cell line. Several 'cell
lines' were further studied after being in culture for more than
six months and had the following properties (6):

a) Morphologically, they resembled flat, simple cuboidal epithelial
cells.

b)   By immunofluorescence, using a monoclonal antibody specific
for epithelial cells (8), keratin patterns typical of epithelial
cells were observed. 100% of the cells were stained, indicating a
loss of fibroblasts during the establishment of the lines.

c) Cells could be sub-cultured at a 1:10 dilution weekly and grown
on plastic in E4 media + 5% foetal calf serum. (Several lines were
kept more than a year and half continuously in culture.) In 1%
foetal calf serum, cells grew normally for several months, then the
growth rate gradually slackened.

d) In soft agar assays (used to measure anchorage-independent
growth), a limited number of rounds of cell division were observed,
after which growth ceased. Cells returned to plastic after two
months in soft agar resumed normal growth.

e) Cell injected into nude (nu/nu) mice failed to produce tumours.

f) 'Footprints' of EB viral DNA could be observed by restriction

enzyme digestion of total chromosomal DNA with *Bam*HI, followed by
fragment separation, transfer to nitrocellulose, and hybridisation
against ^{32}P-labelled viral DNA probes. In early passage cells, many
viral bands were observed, the pattern of which became simpler and
ultimately stabilised with time (5).

The conclusion allowed by these experiments is that EBV probably
contains a gene(s), located in the "right-hand quadrant"of the genome,
that is capable of 'immortalising' primary epithelial cells, that
is, stimulating them to proliferate continuously, at least in culture. That these cells are not 'fully-transformed' is shown by
their limited growth in low serum and semi-solid media, and their
inability to produce tumours in immuno-incompetent animals.

Another animal model was sought for similar studies in order to:

a) Determine whether the epithelial cell immortalisation was a
general property of the viral sub-genomic fragments p13 (p33) and
p31, or was specifically related to (or confined to) the particular
AGMK cells themselves.

b) Find a better model for assessing chromosomal changes.

If specific translocations such as observed with Burkitt's lymphomas
(9, and references therein) are requisite accompanying steps in the
production of a fully-transformed, or malignant, phenotype, then
the difficult chromosome pattern of AGMK cells (diploid number =
60) (10) would complicate chromosomal characterisations. For these
and other reasons (5), fresh kidney cells from common marmosets
(*Callithrix jacchus*, diploid number = 46 (10))were put in culture,
and transfected with fragments of EBV, as described above. Once
again, it was observed that p31, one of the two overlapping fragments that induced immortalisation of AGMK cells, was competent to

immortalise a sub-population of marmoset kidney cells (11). Further
analysis of these cells revealed that in the presence of the EBV
fragment, epithelial cells had been induced to outgrow the fibroblasts in the total population. The properties of the marmoset
cells were similar to those described for the AGMK cells, that is,
they are immortalised but not fully-transformed or malignant. After
a half year in culture, their chromosomal pattern appeared to be
unaltered, and EBV DNA 'footprints' could be detected in total
chromosomal DNA. These cell lines provide materials for further
studies on the EBV DNA involved in the immortalisation event. Also,
they provide ideal models for investigating agents, such as chemical carcinogens, oncogenes, etc., which might induce the further
step(s) needed for full-cellular transformation (4). At the moment,
our efforts are largely concentrated on identifying the viral gene
responsible for the altered cellular phenotype.

In an extension of this study to human cells, it has been shown
that transfection by p31 results in growth alteration in mammary
cells (11). Further, a sub-population of human foetal kidney cells
has been in culture for nearly half a year and are growing well
(our unpublished data). Thus it would appear that a function capable of immortalising epithelial cells from a variety of sources
may reside within the viral genome. It remains to be determined
whether there is any relationship between this function and that
involved in the human disease, NPC, or the B cell malignancy.

EPITHELIAL CELLS VERSUS B-LYMPHOCYTES

The data described above are in apparent conflict with some of the
studies directed towards localising a (so-called) 'transforming'
function in B-lymphocytes. *In vitro* EB virions stimulate B-cells
to grow continuously in culture, although lymphoblastoid lines so-
created do not usually grow in soft agar nor produce tumours in
nude mice (4). Cell lines established from Burkitt's lymphomas, on

the other hand, are tumourigenic and are now known to express not
only EBV genes, but to have the c-*myc* oncogene on chromosome-8
translocated in such a way that its expression apparently comes
under the control of immunoglobulin genes (9). That is to say,
c-*myc* appears to be 'de-regulated'. These data are not inconsistent
with our results obtained from studies on epithelial cells. However,
earlier experiments suggested that the B-cell immortalising function
might be the EB nuclear antigen, EBNA (12), subsequently localised
within the *Bam*HI-K fragment (13); this fragment is not encompassed
within the recombinant DNA cosmid clone, p31, nor has EBNA been
detected in any of our epithelial cell lines. Earlier support for
the notion of EBNA as a transforming function came from the fact
that it is ubiquitously expressed both in Burkitt's and NPC cells
(1). Further, it was originally thought to complex with the host
cell protein, p53 (14) and was postulated thus to play a role in
cellular transformation possibly in a manner analogous to SV40
large T-antigen (15). These experiments have not been held up, however, to further scrutiny.

Another candidate for a transforming gene is a funcion encoded within the *Bam*HI W-Y-H region of the genome (see Figure 1). A 'non-
transforming' EB virus, designated P3HR-1, has been isolated from
a Burkitt's lymphoma line, JIJOYE (16). It has a deletion that
encompasses a portion of one of the large internal repeats (*Bam*HI-W),
all of the *Bam*HI-Y fragment and a part of *Bam*HI-H; this deletion is
generally assumed to be relevant to its loss of ability to immortalise B-cells. Indeed, it has been reported that transforming virus
can be rescued from P3HR-1 cells by complementation with a subgenomic fragment of EBV DNA that covers the deletion (17). These
experiments have not, however, been reproducible in other laboratories (D.J. Volsky, personal communication).

Thus, at this juncture, we are left with the possibility that the immortalising function localised within p31 may still have relevance for B-lymphocytes. On the other hand, there is little doubt that a genome as large as that of EBV (172,000 base pairs) (18) may well have encoded within it more than one immortalising function. The dilemma concerning B-lymphocyte immortalisation will undoubtedly be resolved once direct experiments, similar to those already underway with epithelial cells, have been carried out.

CONCLUSIONS

The data derived from transfection of epithelial cells in culture are consistent with epidemiological evidence (1,2) which suggests that EBV encodes a function that causes continuous cellular pro- liferation. A second event, or co-factor, is necessary for product- ion of a malignant phenotype. At present, therefore, the viral role of EBV in human disease appears to be different from that of another herpes virus (herpes similex) where a 'hit and run' mechanism has been proposed (20). In the long run, isolation of the EBV-encoded function will be required in order to prove definitively this dif- ference.

REFERENCES

1) EPSTEIN, M.A. & ACHONG, B.G. eds. (1979). "The Epstein-Barr Virus", Springer, Berlin.
2) KLEIN, G. ed. (1980). "Viral Oncology", Raven Press, New York.
3) SULLIVAN, J.L., MEDVECZKY, P., FORMAN, S.J., BAKER, S.M., MONROE, J.E. & MULDER, C. (1984). Epstein-Barr-Virus induced lymphoproliferation. New Eng. J. Med., 311, 1163.
4) NILSSON, K., GIOVANELLA, B.C., STEHLIN, J.S. & KLEIN, G. (1977). Tumorigenicity of human hematopoeitic cell lines in athymic nude mice. Int. J. Cancer, 19, 337.
5) GRIFFIN, B.E. & KARRAN, L. (1984). Immortalisation of monkey

epithelial cells by specific fragments of Epstein-Barr virus DNA. Nature, 309, 78.

6) HOHN, B. & COLLINS, J. (1980). A small cosmid for efficient cloning of large DNA fragments. Gene, 11, 291.

7) VAN DER EB, A.S. & GRAHAM, F.L. (1980). Assay of transforming activity of tumor virus DNA. In: "Methods in Enzymology", 65, 826.

8) LANE, E.B. (1982). Monoclonal antibodies provide specific intramolecular markers for study of epithelial tonofilament organisation. J. Cell. Biol., 92, 655.

9) DALLA-FAVERA, R., BREGNI, M., ERIKSON, J., PATTERSON, D., GALLO, R.C. & CROCE, D.M. (1982). Human c-*myc onc* gene is located on the region of chromosome 8 that is translocated in Burkitt's lymphoma cells. Proc. Natl. Acad. Sci. USA, 79, 7824.

10) HSU, T.C. & BENIRSCHKE, K. (1967). "An Atlas of Mammalian Chromosomes", Vol. 1, Folio 1, Springer-Verlag, Berlin, pp. 46, 48.

11) GRIFFIN, B.E., KARRAN, L., KING, D. & CHANG, S.E. (1985). Immortalising gene(s) encoded by Epstein-Barr virus. Soc. Gen. Microbiol., Cambridge Univ. Press, in press.

12) REEDMAN, B.M. & KLEIN, G. (1973). Cellular localization of an Epstein-Barr virus (EBV) associated complement-fixing antigen in producer and non-producer lymphoblastoid cell lines. Int. J. Cancer, 11, 499.

13) SUMMERS, W.P., GROGAN, E.A., SHEDD, D., ROBERT, M., LIU, C.-R. & MILLER, G. (1982). Stable expression in mouse cells of nuclear neoantigen after transfer of a 3.4-megadalton cloned fragment of Epstein-Barr virus DNA. Proc. Natl. Acad. Sci. USA, 79, 5688.

14) LUKA, J., JORNVALL, H. & KLEIN, G. (1980). Purification and biochemical characterization of the Epstein-Barr virus-determined nuclear antigen and an associated protein with a 53,000 dalton sub-unit. J. Virol., 35, 592; STRNAD, B.C., SCHUSTER, T.C., HOPKINS, R.F., NEUBAUER, R.H. & RABIN, H. (1981). Identification of an Epstein-Barr virus nuclear antigen by fluor-

immunoelectrophoresis and radioimmunoelectrophoresis. *ibid*, 38, 966.

15) LANE, D.P. & CRAWFORD, L.V. (1979). T-antigen is bound to a host protein in SV40-transformed cells. Nature, 278, 261; LINZER, D.I.H. & LEVINE, A.J. (1979). Characterization of a 54K dalton cellular SV40 tumor antigen present in SV40-transformed cells and uninfected embryomal carcinoma cells. Cell, 17, 43.

16) HINUMA, Y., KOHN, M., YAMAGUCHI, J., WUDARSKI, J., BLAKESLEE, J.R. & GRACE, J.T. (1967). Immunofluorescence and herpes-type virus particles in the P3HR-1 Burkitt lymphoma cell line. J. Virol., 1, 1045; RABSON, M., GRADOVILLE, L., HESTON, L. & MILLER, G. (1982). Non-immortalizing P3J-HR-1 Epstein-Barr virus; a deletion mutant of its transforming parent Jijoye. *ibid*, 1, 44, 834.

17) STOERKER, J. & GLASER, R. (1983). Rescue of transforming Epstein-Barr virus (EBV) from EBV-genome-positive epithelial hybrid cells transfected with subgenomic fragments of EBV DNA. Proc. Natl. Acad. Sci. USA, 80, 1726.

18) BAER, R., BANKIER, A.T., BIGGIN, M.D., DEININGER, P.L., FARRELL, P.J., GIBSON, T.G., HATFULL, G., HUDSON, G.S., SATCHWELL, S.C., SEGUIN, C., TUFFNELL, P.S. & BARRELL, B.G. (1984). DNA sequence and expression of the B95-8 Epstein-Barr virus genome. Nature, 310, 207.

19) GIVEN, D. & KIEFF, E. (1978). DNA of Epstein-Barr virus, IV. Linkage map of restriction enzyme fragments of the B95-8 and W91 strains of Epstein-Barr virus. J. Virol., 28, 524; SKARE, J. & STROMINGER, J.L. (1980). Cloning and mapping of *Bam*HI endonuclease fragments of DNA from the transforming B95-8 strain of Epstein-Barr virus. Proc. Natl. Acad. Sci. USA, 77, 3869.

20) GALLOWAY, D.A. & McDOUGALL, J.K. (1983). The oncogenic potential of herpes simplex viruses: evidence for a 'hit and run' mechanism. Nature, 302, 21.

FUNCTIONAL DOMAINS OF PURIFIED ADENOVIRUS TYPE C E1A PROTEINS

B. Krippl*, B. Ferguson+, N. Jones#, M. Rosenberg+ &
H. Westphal*

*Laboratory of Molecular Genetics, National Institute of
Child Health and Human Development, National Institutes
of Health, Bethesda, Maryland, USA; +Department of Mole-
cular Genetics, Smith Kline and French Laboratories
Philadelphia, Pennsylvania, USA; #Purdue University
West Lafayette, Indiana, USA

INTRODUCTION

The adenovirus E1A gene controls initial stages of oncogenic transformation (1-5) and activates transcription of a variety of viral
and host cell genes (6-11). We have a strong interest in the molecular details of these important regulatory functions. For this
reason, some of us (12) have inserted the E1A coding sequence in a
prokaryotic expression vector in order to generate the quantities
of pure E1A proteins that are required to study their action. Here,
we summarize results, to be published in detail elsewhere, which
have allowed us to define more closely those sequences of the E1A
gene that determine the intracellular location of E1A proteins as
well as their functions during the virus life cycle.

E1A PROTEINS MADE IN *E. coli*

We inserted E1A coding sequences into the plasmid vector pAS1
which has been designed to produce foreign proteins at high yields
in *E. coli*. Details of our constructs have been published (12).

The first E1A protein produced by this approach was the 289 amino-
acid product of the 13S mRNA of serotype C adenovirus (13,14).
This protein was synthesized in large amounts in *E. coli* (12) and
has since been purified to near homogeneity, utilizing the properties that it remained insoluble in the presence of high salt
concentrations, but could be solubilized in the presence of dena-
turing agents to yield a highly active preparation (15). By de-
leting one or the other part of the E1A, 13S mRNA coding sequence
from the pAS1 vector, we have also generated a number of truncated
proteins (unpublished). Their amino acid compositions are described below. Finally, we have recently generated a facsimile of
the 12S E1A mRNA product, a protein that, like the 289 amino-acid,
13S mRNA product, is found in the infected cell, but differs from
the latter by the lack of an internal sequence of 46 amino acids
(13,14).

FUNCTIONAL ACTIVITIES OF E1A PRODUCTS IN THE MAMMALIAN CELLS

E1A functional activities were determined after microinjection of
purified proteins into Vero cells, a monkey kidney cell line permissive for adenovirus growth. In each case, we asked whether the
protein, when injected into the cytoplasm, would move to the nucleus, and whether it would complement H5dl312, a deletion mutant
of adenovirus which is unable to produce E1A functions by itself
(7). Details of our experimental procedures may be found elsewhere
(15). In short, nuclear movement was monitored by fluorescent
staining of cells at various intervals after protein injection into the cytoplasm. For the dl312 complementation test, cells were
infected with the mutant and injected with protein. After an appropriate incubation time, cells were pulse labeled with [S35]
methionine, and viral coat proteins were immunoprecipitated and
separated by SDS-PAGE. The coat protein signal served as a numerical indicator of the degree of complementation.

Table 1. Functions of E1A proteins

| Protein | Amino acid composition | d1312 effect when injected at | | | Nuclear movement |
		0.2 mg/ml	0.5 mg/ml	1 mg/ml	
13	1+3-289	12	50	100	fast
12	1+3-139+186-289	11	45	58	fast
13-T1	1+3-223	7	46	67	slow
13-T2	1+22-223	0	12	54	slow
13-T3	1+22-289	0	18	64	fast
13-T4	1+3-152(+18)	0	0	18	slow
13-T5	1+3-120(+48)	0	0	0	slow

The results are summarized in Table 1. The principal protein, termed 13, is the product of the 13S mRNA. Because of the way the 13S cDNA was inserted into the expression vector, the second of the 289 amino acids of this protein is missing from the product made in *E. coli*. This is indicated under the heading "amino acid composition" by the formula 1+3-289. Protein 12, the product of 12S mRNA, lacks an additional internal sequence of 46 amino acids, from position 140 to 185, hence its formula 1+3-139+186-289. All other proteins, 13-T1 through 13-T5, are products of partially deleted E1A sequences, therefore representing truncated forms of protein 13. Because of the way the expression vectors were constructed, proteins 13-T4 and 13-T5 carry missense carboxy terminal sequences, 18 and 48 amino acids in length, respectively.

The d1312 complementation test was performed by injecting protein at a concentration of 0.2, 0.5 or 1 mg/ml and measuring the amount of d1312 coat protein produced in the injected cells. Table 1 lists relative values, the maximal value being 100. Our results show that, of all proteins tested, only 13-T5 fails to complement dl312, whereas protein 12 and most of the truncated derivatives of protein 13 retain considerable complementation activity.

Also noted in Table 1 is the nuclear movement of the various E1A proteins. Nuclear localization was found to be either fast and completed within minutes, or slow and incomplete even after many hours of incubation, depending on the type of protein injected. Slow distribution of a protein throughout the cell may actually reflect a passive diffusion process.

Two additional properties, not listed in the table, were determined for protein 13. (a) The protein has a remarkable longevity in vivo. Cells injected with the protein up to 18 hours prior to dl312 infection still were able to express dl312 coat proteins. (b) The protein is extremely heat stable. Heating at 100^{0}C for 5 minutes did not abolish dl312 complementation.

CONCLUSIONS

Our results demonstrate that the *E. coli* made E1A proteins exhibit all the properties of the genuine viral regulatory gene necessary to complement a virus defective in that gene. This complementation is tightly coupled with nuclear localization. Nuclear movement, therefore, appears to be an intrinsic property of a functional E1A product.

Proteins 13 and 12 both complement dl312. On a quantitative basis, the former appears to be somewhat more efficient than the latter. In a similar experiment, Winberg and Shenk (16) generated mutants which carried either 13S cDNA or 12S cDNA in the place of the normal E1A gene. Both viruses grew well in HeLa cells, but titers were higher with the mutant producing the 13S mRNA product. The fact that both the 13S mRNA product and the 12S mRNA product are expressed in cells productively infected or transformed by wild type C adenovirus suggests distinct roles for each protein. Which, then, are their unique properties? An answer to this question may come from a thorough analysis of the physical properties of each protein, including their post translational modifications. Equally important in this regard is the identification of cellular and viral proteins and nucleic acids with which they interact. Experiments along these lines are currently in progress.

Also in progress is our work with truncated E1A proteins. From our present experiments, we can conclude that removal of a limited number of amino acid residues from the amino and carboxy termini of protein 13 does not severely impede its role in dl312 complementation. By contrast, nuclear movement appears to be curtailed by removal of carboxy terminal sequences. Additional constructs are required to define precisely the various functional domains of E1A. There is every reason to believe that this important eukaryotic regulatory gene will provide excitement for quite some time to come.

REFERENCES

1) HOUWELING, A., VAN DEN ELSEN, P.J., & VAN DER EB, A. (1980). Partial transformation of primary rat cells by the leftmost 4.5% fragment of adenovirus 5 DNA. Virology, 105, 537.

2) RULEY, H.E. (1983). Adenovirus early region 1A enables viral and cellular transforming genes to transform primary cells in culture. Nature, 304, 602.

3) VAN DEN ELSEN, P., DE PATER, S., HOUWELING, A., VAN DER VEER, J., & VAN DER EB, A. (1982). The relationship between region E1a and E1b of human adenoviruses in cell transformation. Gene, 18, 175.

4) BERNARDS, R., SCHRIER, P.I., BOS, J.L., & VAN DER EB, A.J. (1983). Role of adenovirus types 5 and 12 early region 1b tumor antigens in oncogenic transformation. Virology, 127, 45.

5) VAN DEN ELSEN, P.J., HOUWELING, A., & VAN DER EB, A.J. (1983). Morphological transformation of human adenoviruses is determined to a large extent by gene products of region E1a. Virology, 131, 242.

6) BERK, A.J., LEE, F., HARRISON, T., WILLIAMS, J., & SHARP, P.A. (1979). Pre-early adenovirus 5 gene product regulates synthesis of early viral messenger RNAs. Cell, 17, 935.

7) JONES, N. & SHENK, T. (1979). An adenovirus type 5 early gene function regulates expression of other early viral genes. Proc. Natl. Acad. Sci. USA, 76, 3665.

8) NEVINS, J.R. (1981). Mechanism of activation of early viral transcription by the adenovirus E1A gene product. Cell, 26, 213.

9) WEEKS, D.L. & JONES, N.C. (1983). E1A control of gene expression is mediated by sequences 5' to the transcriptional starts of the early viral genes. Mol. Cell. Biol., 3, 1222.

10) GREEN, M.R., TREISMAN, R., & MANIATIS, T. (1983). Transcriptional activation of cloned human beta-globin genes by viral immediate-early gene products. Cell, 35, 137.

11) LEFF, T., ELKAIM, R., GODING, C.R., JALINOT, P., SASSONE-CORSI, P., PERRICAUDET, M., KÉDINGER, C., & CHAMBON, P. (1984). Individual products of the adenovirus 12S and 13S E1a mRNAs stimulate viral EIIa and EIII expression at the transcriptional level. Proc. Natl. Acad. Sci. USA, 81, 4381.

12) FERGUSON, B., JONES, N., RICHTER, J., & ROSENBERG, M. (1984). Adenovirus E1a gene product expressed at high levels in *Escherichia coli* is functional. Science, 224, 1343.

13) PERRICAUDET, M., AKUSJÄRVI, G., VIRTANEN, A., & PETTERSSON, U. (1979). Structure of two spliced mRNAs from the transforming region of human subgroup C adenoviruses. Nature, 281, 694.

14) BAKER, C.C. & ZIFF, E.B. (1981). Promoters and heterogeneous 5' termini of the messenger RNAs of adenovirus serotype 2. J. Mol. Biol., 149, 189.

15) KRIPPL, B., FERGUSON, B., ROSENBERG, M., & WESTPHAL, H. Func-
    tions of purified E1A protein microinjected into mammalian
    cells. Proc. Natl. Acad. Sci USA, in press.
16) WINBERG, G. & SHENK, T. (1984). Dissection of overlapping
    functions within the adenovirus type 5 E1A gene. EMBO J., 3,
    1907.

PARVOVIRUSES AND CANCER

Bernhard Hirt

Swiss Institute for Experimental Cancer Research
Ch. des Boveresses, 1066 Epalinges, Switzerland

Parvoviruses are unique among the animal viruses by the fact that their genome consists of single stranded DNA. They are also the only group of DNA viruses which so far does not include any tumor viruses.

You have heard during this course about the different functions of polyoma viruses, SV40 and adenoviruses which lead to cell transformation. In transformed cells the viral DNA is covalently integrated into the host genome. The techniques developed to study polyoma and SV40 proved very useful for the analysis of two types of small DNA viruses which are closely linked to frequent human cancers, and which both so far cannot be grown in tissue culture. I think that it is useful to mention this here, because these are very important and recent findings which are not covered by the other lecturers. Thanks to the DNA cloning techniques it became possible to study their implication in human cancer. They are the hepatitis B virus, whose DNA is found in primary carcinoma of the liver (for references see 1), a frequent disease in some African countries and in China, and the human papilloma viruses 16 and 18, whose genomes are detected in a majority of cancers of the cervix

175

of the uterus (2,3). Together with the human T cell leukemia viruses
and the Epstein-Barr virus, these small DNA viruses are already
known viruses implied in human cancer. I am convinced that several
human tumor viruses remain to be discovered.

Let us turn now to the subject of my presentation, the parvoviruses.
We started to study parvoviruses a few years ago because there were
some reports that they might inhibit the appearance of tumors and
also because several parvoviruses have an immunosuppressive effect
and by this might influence the development of cancers.

As already mentioned, the parvoviruses have a genome consisting of
single stranded DNA and its size is about 5 kilobases. Both ends of
the DNA are folded back and form base-paired "hairpin" structures.
The capsid of the virus is built of three proteins and its diameter
is about 22 nanometers. The parvoviruses are not enveloped. They
are among the smallest viruses known. A very characteristic property
of the parvoviruses is that their replication is restricted to
dividing cells, more precisely to cells synthesizing DNA (S-phase).
The parvoviruses do not have the capability to stimulate quiescent
cells to enter the cell cycle, as for instance polyoma virus does
(For reviews see refs. 4 and 5).

The replicating parvoviruses lead to the lysis of the host cell.
This, together with the fact that parvoviruses attack only dividing
cells, is of importance if one considers their pathological effects.

Among the parvoviruses one distinguishes between the autonomous
viruses and the dependoviruses. The latter genus replicates only
in cells which are simultaneously infected with either adenoviruses
or herpesviruses.

AUTONOMOUS PARVOVIRUSES

Those parvoviruses known to cause disease belong to the autonomously replicating genus. Examples are the feline panleukopenia virus,
the porcine parvovirus and the newly detected human parvovirus (6).

In our laboratory we work with the minute virus of mice (MVM), an
autonomous parvovirus which to our knowledge does not cause any
disease. In a first study we tried to obtain cell transformation by
this virus. We infected cultures of permissive murine cells (fibroblasts) with high titres of virus. The infection killed most cells,
but selected a few resistant cells that continued to grow. Virus
could be found in these cultures during up to 25 cell passages
after the infection. Attempts to find integrated viral DNA by the
blot technique of Southern (7) were all negative. All evidence
indicates that the virus has only a lytic function and that it does
not give the infected cells a growth advantage.

We also analyzed a variant of MVM that does not grow on fibroblasts,
but on T-lymphocytes. It is immunosuppressive in mixed leukocyte
cultures and is named MVMi (8,9). DNA sequence analysis shows that
MVM (10) and MVMi (11) are closely related: only at 3.5% of the
nucleotides were differences observed. The question we asked is
whether MVMi would be immunosuppressive *in vivo* and in this way
prevent tumor rejection. This was tested in a model of transplanted
murine tumors. MVM prototype virus, which has no effect on mixed
lymphocyte cultures was injected into control animals. To our surprise, both viruses accelerated the rejection of the graft. This
is illustrated in Figure 1 (P. Kimsey & H. Engers, unpublished
results). We would have expected the MVM prototype to have no influence on the rejection, since it has no effect on mixed leukocyte
cultures. That the MVMi, which *in vitro* kills T-lymphocytes, favors
graft rejection is puzzling and we have no explanation to offer.

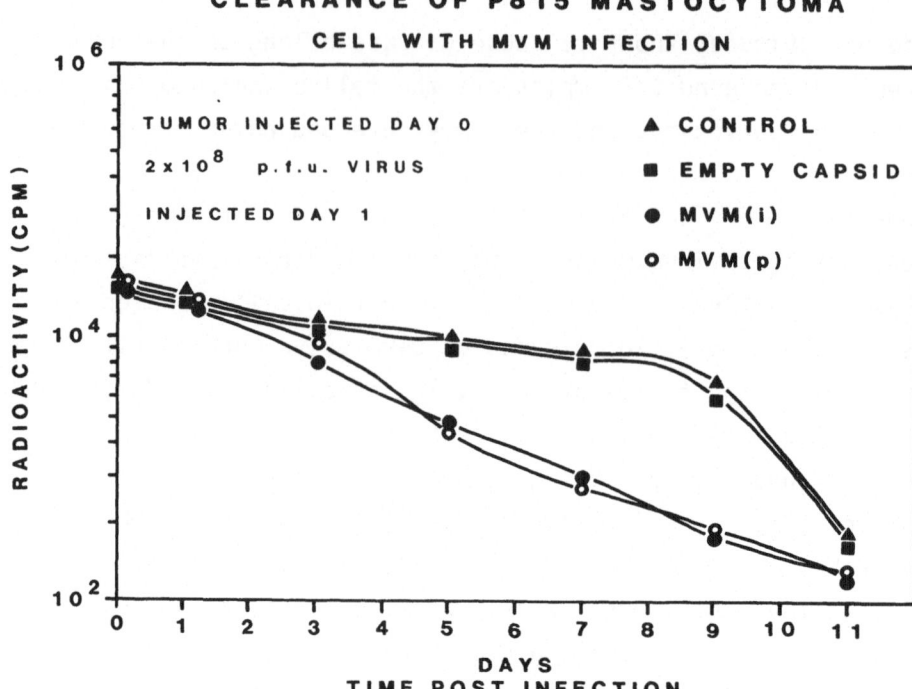

Figure 1. *Rejection of transplanted tumor cells.*

P 815 mastocytoma cells were labelled with radioactive IUdR.
They were injected into C57Bl/6 female mice of approximate age of
2 months. Total body radioactivity was measured to study the
disappearance of radioactive cells. The upper two curves are without virus, the lower two curves in the presence of the indicated
murine parvoviruses.

REDUCTION OF TUMORS

Toolan has made an extensive study on the action of the autonomous
parvovirus H1 of the hamster (12). She wanted to find out whether
it caused tumors and injected it into 1,700 newborn hamsters and
screened them for tumors 2 to 3 years later. The frequency of tumors
observed was about 25 times lower than in the control animals. This
was a very surprising result. In experiments, in which adenovirus
12 (13) and dimethylbenzathracene were used (14) as carcinogens,

infection by H1 also reduced tumor incidence. Infection by parvovirus H1 therefore decreases the appearance of tumors drastically.
Infection of young animals could be a preventive measure in order
to reduce tumor incidence. But the price to be paid is too high,
since hamsters injected with H1 a short time after birth do not
develop normally, but are dwarfed and deformed (12).

The effect of parvovirus infection on cell transformation by SV40
has been studied by Mousset & Rommelaere (15). For this purpose
they selected mouse fibroblasts that were resistant to lysis by MVM.
The frequency of transformation of these cells by SV40 was measured
by colony formation in soft agar. Superinfection of the cells with
MVM reduced the transformation frequency by 90%. The authors also
studied the effect of MVM infection on SV40 transformed cells and
found out that a large proportion of transformed cells were lysed
by MVM, while the untransformed parent cells were completely resistant. This could therefore also explain the reduced transformation
frequency. The authors conclude that the SV40 transformed cells are
more permissive to MVM growth and to cell lysis than the untransformed cells.

If this *in vitro* model represents the effects of parvoviruses on
tumor formation *in vivo*, it would state that tumor cells are preferentially lysed by parvoviruses. Further work is needed to test
the value of this model.

ADENO-ASSOCIATED VIRUSES (AAV)

Adeno-associated viruses only grow in cells which are simultaneously
infected by either adenoviruses or herpes viruses. By themselves
they are not lytic and do not produce any pathogenic effect. Infection by AAV inhibits the replication of adenoviruses (16). This has
as a consequence the reduction of cell lysis and of pathogenicity.

But AAV also inhibits the formation of tumors in hamsters by adenoviruses (17). Since the tumor induction takes place in cells that
do not support the replication of adenovirus DNA, the inhibitory
effect of AAV cannot be the same as in lytic infection. The mechanism of inhibition of transformation is also different from the
one observed by MVM on mouse cells since hamster cells are not
lysed by AAV or/and adenoviruses.

AAV can be shown to reduce the cloning efficiency of adeno transformed cells. This can explain the reduction of tumor formation by
adeno infection. This effect of AAV may be mediated through a reduction of the expression of the adeno tumor antigen (18).

The inhibition of adeno replication has probably a favorable effect
on the host and therefore a biological significance. That AAV on
the other hand inhibits the induction of tumors by adeno is only
of importance to the laboratory, since under natural conditions
adenoviruses are not known to cause tumors, regardless whether AAV
is present or not.

The interaction of AAV DNA with its host cells has been studied in
the absence of the helper adenovirus by Cheung and coworkers (19).
They have injected human cells with high titres of AAV and they
grew clones of these cells for many passages. Using the blot technique of Southern (7) they analyzed the state of the viral genome
in these cells and they also tested whether the cells were able to
produce AAV when superinfected with adenovirus. A considerable
proportion of the cell clones analyzed contained AAV DNA, but no
AAV antigens could be detected. Superinfection with adenovirus lead
to the production of AAV. After 10 passages, AAV DNA was found only
in a state of covalent linkage to the DNA of the host cell, while

at later passages both integrated and free viral DNA were found.
From these analyzes it becomes clear that the DNA of AAV can be
integrated into the DNA of the host and stay there and divide under
the control of the host cell. As far as known, this integration does
not give any growth advantage to the cell. Therefore no selective
growth conditions exist for cells containing AAV DNA. It is interesting that the virus can be reactivated by the superinfection with
adenovirus. The mechanism of the excision of the viral DNA from the
host cell is not known.

It is obvious that AAV has several properties in common with SV40:
its DNA can be integrated into the host and replicate there like a
host gene for many generations. Through a signal from outside the
cell, the virus can be reactivated. But there is a very important
difference. SV40 genes of the integrated DNA are expressed and lead
to an abnormal cell growth. The AAV genes are silent and do not
produce tumors.

The integration of its DNA into the host DNA and its replication as
a non-expressed provirus gives the AAV a possibility to survive in
the absence of its helper virus.

––––––––––

All observations made so far on the relations between parvoviruses
and cancers indicate that the viruses inhibit the appearance of
cancers. More studies are needed to elucidate the mechanisms of
these inhibitions. They are likely to increase our understanding
of tumor formation.

The support of the Fonds National Suisse de la Recherche Scientifique (Grant no. 3.261.82) is acknowledged.

REFERENCES

1) SHAUL, Y.,ZIEMER, M.,GARCIA, P.D.,CRAWFORD, R., HSU, H.
   VALENZUELA, P. & RUTTER, W.J. (1984). Cloning and analysis of
   integrated hepatitis virus sequences from human hepatoma cell
   line. J. Virol., 51, 776-787.

2) BOSHART, M., GISSMANN, L.,IKENBERG, H., KLEINHEINZ, A.
   SCHEURLEN, W. & ZUR HAUSEN, H. (1984). A new type of papillomavirus DNA, its presence in genital cancer biopsies and in cell
   lines derived from cervical cancer. EMBO J., 3, 1151-1157.

3) CRAWFORD, L. (1984). Papilloma viruses and cervical tumours.
   Nature, 310, 16.

4) WARD, D.C. & TATTERSALL, P., eds., (1978). "Replication of
   Mammalian Parvoviruses", Cold Spring Harbor Laboratory.

5) BERNS, K.I., ed. (1984). "The Parvoviruses", Plenum Press.

6) ANDERSON, M.J., JONES, S.E., FISCHER-HOCH, S.P., LEWIS, E.,
   HALL, S.M., BARTLETT, C.L., COHEN, B.J., MORTIMER, P.P. &
   PEREIRA, S. (1983). Human parvovirus, the cause of erythema
   infectiosum (fifth disease)? The Lancet, June 18, 1378.

7) SOUTHERN, E. (1975). Detection of specific sequences among DNA
   fragments separated by gel electrophoresis. J. Mol. Biol., 98,
   503-518.

8) McMASTER, G.K., BEARD, P., ENGERS, H.D. & HIRT, B. (1981).
   Charcterization of an immunosuppressive parvovirus related to
   the minute virus of mice. J. Virol., 38, 317-326.

9) ENGERS, H.D., LOUIS, J.A., ZUBLER, R.H. & HIRT, B. (1981).
   Inhibition of T-cell mediated functions by MVM(i), a parvovirus
   closely related to the minute virus of mice. J. Immmunol., 127,
   2280-2285.

10) ASTELL, C.R., THOMSON, M., MERCHLINSKY, M. & WARD, D.C. (1983).
    The complete DNA sequence of minute virus of mice, an autono-
    mous parvovirus. Nucl. Acids Res., 11, No. 4, 999-1018.

11) SAHLI, R. (1985). The nucleotide sequence of the immunosuppress-

ive variant of minute virus of mice. Ph.D. Thesis, University of Lausanne.

12) TOOLAN, H.W. (1967). Lack of oncogenic effect of the H-viruses for hamsters. Nature, 214, 1036.

13) TOOLAN, H.W. & LEDINKO. N. (1968). Inhibition by H-1 virus of the incidence of tumors produced by adenovirus 12 in hamsters. Virol., 35, 475-478.

14) TOOLAN, W., RHODE, S.L., III & GIERTHY, J.F. (1982). Inhibition of 7,12-dimethylbenz(a)anthracene-induced tumors in Syrian hamsters by prior infection with H-1 parvovirus. Cancer Res., 42, 2552-2555.

15) MOUSSET, S. & ROMMELAERE, J. (1982). Minute virus of mice inhibits cell transformation by simian virus 40, Nature, 300, 537-539.

16) HOGGAN, M.D., BLACKLOW, N.R. & ROWE, W.P. (1966). Studies of small DNA viruses found in various adenovirus preparations: Physical and immunological characteristics, Proc. Natl. Acad. Sci. USA, 55, 1467-1471.

17) KIRCHSTEIN, R.L., SMITH, K.O. & PETERS, E.A. (1968). Inhibition of adenovirus-12 oncogenicity by adeno-associated virus. Proc. Soc. Exp. Biol. Med., 128, 670-673.

18) OSTROVE, J.M., DUCKWORTH, D.H. & BERNS, K.I. (1981). Inhibition of adenovirus-transformed cell oncogenicity by adeno-associated virus, Virology, 113, 521-533.

19) CHEUNG, A.K.M., HOGGAN, M.D., HAUSWIRTH, W.W. & BERNS, K.I. (1980). Integration of the adeno-associated virus genome into cellular DNA in latently infected human Detroit 6 cells. J. Virol., 33, 739-748.

HTLV IN ADULT T CELL LEUKEMIA AND ACQUIRED IMMUNE DEFICIENCY SYNDROME

Prem.S. Sarin

Laboratory of Tumor Cell Biology, National Cancer
Institute, Bethesda, Maryland 20205, USA

INTRODUCTION

The involvement of retroviruses in animal leukemias and lymphomas
has been well established. Retroviruses have been isolated from
several animal species including chickens, mice, cats, cows and
gibbon apes (1). Leukemia induced by retroviruses in mice and cats
is usually accompanied by viremia and abundant virus production.
Two animal model systems, *viz.*, the feline and bovine leukemias
provide some interesting parallels with the human disease. In the
feline system, the isolation of a retrovirus (FeLV) from leukemic
cats, the epidemiology of the disease and induction of leukemia by
this virus clearly established FeLV as the cause of the disease in
cats (2,3). In the case of bovine leukosis, a disease found to cluster in cattle herds, the involvement of a virus in the causation
of the disease was considered long before the isolation of bovine
leukemia virus (BLV). BLV was not detected in primary leukemic cells
but was expressed in cells only after short term culture. BLV has
since been isolated and transmitted to permanently growing cell
lines and fully characterized (4). Early attempts to isolate a
retrovirus from human leukemic tissues were unsuccessful until the
availability of techniques to grow T cells in culture and the recent

185

isolation of human T cell leukemia virus (HTLV) from patients with
adult T cell malignancies and acquired immune deficiency syndrome
(AIDS).

Retroviruses can be either transmitted vertically (endogenous) in
the germ line or are acquired by horizontal (exogenous) infection.
No naturally occurring malignancy has been shown to be associated
with endogenous retroviruses in animal or human systems. Endogenous
retroviruses have, however, been detected in a number of animal
species and in human DNA (5). Exogenous retroviruses containing an
onc gene (acutely transforming viruses) are usually defective because they have lost some genetic information for virus replication
during the acquisition of an *onc* gene (6). For the development of
malignancy, usually a sarcoma or acute leukemia, with an *onc* gene
containing virus, the presence of a nondefective helper retrovirus
is essential. The chronic leukemia viruses, which lack an *onc* gene,
have been associated with naturally occurring leukemias and lymphomas (1), are replication competent, and require a long incubation
period to induce malignancy. Although the molecular mechanisms by
which chronic leukemia viruses induce disease are not clear, studies
in the chicken system indicate that activation of the c-*myc* gene by
integration of the provirus in its vicinity may be important for
the induction of avian leukosis (7).

ISOLATION OF HUMAN T CELL LEUKEMIA VIRUS (HTLV)

The availability of a growth factor called the T cell growth factor
(TCGF, IL-2) (8,9) made it possible to grow T cells in culture from
patients with T cell malignancies and to examine them for retrovirus expression. The first human retrovirus (HTLV) was subsequently detected in T cell cultures from a patient with adult T cell
leukemia (10). The U.S. patients from which HTLV was first isolated
were considered to be suffering from variants of mycosis fungoides
and Sezary syndrome (10,11), but a reexamination of the clinical

history of these patients showed that their clinical features were identical to a disease commonly described as adult T cell leukemia/ lymphoma (ATLL) in Japan (12,13) or T lymphosarcoma cell leukemia (TLCL) in Caribbean Black immigrants in London (14). HTLV is a type-C retrovirus (100 nm), buds from the cell membrane, contains reverse transcriptase (90,000 daltons) (15,16), high molecular weight RNA (16,17), and envelope and internal core proteins (p19,p24)(18,19). Over 100 new HTLV isolates have since been obtained from patients in the U.S., the Caribbean, Japan, Africa, South America, U.K., and Israel (16,20-25). These HTLV isolates have been obtained from cultured T cells of ATLL patients and their normal family members (26).

Table 1. *Representative examples of HTLV isolates belonging to subgroups 1, 2 and 3 (16,20)*

Designation	Cell Line	Geographic Origin	RNA	p24	p19
A. HTLV-1					
CR	HUT102	U.S.	+++	+++	+++
MJ	MJ	U.S.	++	++	++
MB	MB	Caribbean	++	++	++
MI	MI	Caribbean	++	++	++
UK	UK	Israel	++	++	++
SK	SK	Japan	ND	++	++
SD	SD	Japan	++	++	++
MT-2	MT-2	Japan	++	++	++
B. HTLV-2	MO	U.S.	±	+	−
C. HTLV-3	H9V	U.S.	±	±	−

++ = Highly related to or indistinguishable from the first HTLV isolate (HTLV-1$_{CR}$); ± = Slightly related to HTLV-1$_{CR}$; + = Related but readily distinguishable from HTLV-1$_{CR}$.

HTLV isolates obtained from these patients are related to one another and have been grouped into HTLV subgroup 1 (Table 1). These
HTLV isolates show similar immunological cross-reaction of the
viral proteins, sequence homology by molecular hybridization, and
cleavage sites of several restriction endonucleases to the first
HTLV-1 isolate from a black U.S. patient (HTLV-1$_{CR}$). The cytogenetic and HLA patterns of the cultured T cell lines match the characteristic patterns of the fresh donor cells (11,16,20). HTLV isolates, belonging to subgroups 1, 2, and 3,have been isolated from
patients with T cell malignancies (HTLV-1), hairy cell leukemia
(HTLV-2) and acquired immune deficiency syndrome (AIDS) (HTLV-3).
These HTLV isolates are T cell tropic, infect T helper cells
(OKT4+) and the biochemical properties of reverse transcriptase
are similar to the properties of the reverse transcriptase from
the first HTLV isolate.

HTLV SUBGROUPS 2 AND 3

Virus isolates belonging to HTLV subgroup 2 have been obtained (27)
from a patient (MO) with a T cell variant of hairy cell leukemia
(28) and a patient with AIDS (29). A comparison of the core proteins and nucleic acid sequences of HTLV-2 and HTLV-1 showed that
the two viruses were similar but not identical (30). Examination of
the HTLV-1 and HTLV-2 isolates by the techniques of syncytia formation and the HTLV-vesicular stomatitis virus pseudotypes also showed that the HTLV isolates belonging to subgroups 1 and 2 had markedly different envelope antigens (31,32). HTLV isolates belonging
to subgroup 3 have also been isolated and characterized from a
number of patients with AIDS. A T cell line, H9V, has been productively infected with HTLV-3 and produces high titers of this virus.
All HTLV isolates belonging to subgroups 1, 2, and 3 have a Mg2^{+}
dependent reverse transcriptase and are T cell tropic (OKT4 positive
T helper cells).

BIOCHEMICAL PROPERTIES OF HTLV

Molecular hybridization studies utilizing HTLV-1 prototype, HTLV-1 $_{CR}$, showed that HTLV-1 was not related to a wide variety of previously described animal retroviruses (16,17). Similarly, the viral *gag* proteins p24 (18) and p19 (19), and the viral reverse transcriptase (15) were also found to be unrelated to similar proteins from known animal retroviruses. HTLV is T cell tropic and is acquired by exogenous infection. Studies with HTLV and a Japanese virus isolate called ATLV, using Southern blotting technique, show that HTLV is not an endogenous virus and is present in T cells of infected patients (16,22,48). HTLV-1 and "ATLV" have been shown to be closely related by molecular hybridization and by antigenic cross-reaction of the *gag* proteins (33). The HTLV isolates that are closely related by nucleic acid hybridization and antigenic cross-reaction to HTLV-1 $_{CR}$ have been classified as HTLV type 1.

The core proteins of HTLV-1 that have been well characterized include p19, p24, p15, and reverse transcriptase. The reverse transcriptase is a protein of ~100,000 daltons, prefers Mg^{++} as the divalent cation (15,16) and shows no antigenic cross-reaction with reverse transcriptases from other avian and mammalian retroviruses including BLV. HTLV-1 p24 and p15 are also antigenically unrelated to proteins from other mammalian retroviruses (18,19). HTLV-1 p24 shows antigenic cross-reactivity with HTLV-2 p24 (27), but the p19 or p15 from HTLV-1 and HTLV-2 are distinct. Amino acid sequence analyses of HTLV p24 (34) and p15 (35) show a distant relatedness of HTLV-1 proteins with BLV proteins.

SEROEPIDEMIOLOGY OF HTLV

Analyses of the sera from patients with leukemias and lymphomas for HTLV-1 antibodies show that the sera from patients with non-T cell leukemias and lymphomas and from patients with T cell malig-

Table 2. *Incidence of HTLV-1 antibodies in sera of patients with leukemia-lymphoma* (36,38)

Disease Category[a]	# Positive/# Tested	% Positive
ATLL/T-LCL	55/62	89
T-ALL	0/18	0
T-CLL	0/40	0
CTCL	6/260	2
T Hairy Cell Leukemia	1/1	
T-NHL	25/68	37
NHL (non-T)	0/54	0
ALL (non-T)	6/167	4
CLL (non-T)	4/28	1
Hodgkins Lymphoma	0/120	0

[a]ATTL/T-LCL, adult T cell leukemia, T lymphosarcoma cell leukemia; T-NHL, nonHodgkins lymphoma; T-CLL, T chronic lymphocytic leukemia; T-ALL, adult lymphocytic leukemia T cell type; CTCL, cutaneous T cell lymphoma. The patients examined included patients from the United States, Japan and the Caribbean. The incidence of HTLV-1 antibodies in healthy individuals in the endemic regions of Japan, such as Kyushu and Shikoku is 10-16%; in the Caribbean the incidence is 6-12%; in the United States, especially Georgia and Florida, the incidence is 1-2% whereas in West Germany, Netherlands and the United Kingdom no HTLV-1 infection was detectable in healthy individuals. HTLV-1 infection was also observed in Surinam immigrants to Holland (12%), different parts of Venezuela (1-14%), and Africa (2-8%).

nancies which are not of the typical HTLV type are usually negative while sera from the typical HTLV-associated adult T cell malignancies are positive in 90% of the cases (Table 2) (36). These studies show that HTLV is endemic in southwestern parts of Japan, the Caribbean, Africa, and South and Central America. Nearly all the Japanese patients and the West Indian ATLL patients living in England have HTLV antibodies in their serum and these antibodies have also been found in about 70% of all the non-Hodgkin's lymphomas seen in Jamaica (37). Whether HTLV plays any role in the induction of other adult T cell malignancies such as cutaneous T cell lymphomas and leukemias or mycosis fungoides and Sezary syndrome is still unresolved, and analysis of these sera for HTLV antibodies in the past have mostly been negative (38). Recently, with the use of more sensitive ELISA assays it has been found that mycosis fungoides from Denmark have low levels of HTLV antibodies in 15% of the patients tested (39).

Examination of sera from normal donors for HTLV antibodies from the United States and Europe generally shows less than 1% positive cases (38). The normal Japanese population in areas endemic for ATLL has a high proportion of HTLV antibodies (16%) compared to the non-endemic areas (0-2%) (36,40,41). Analysis of the sera for HTVL-1 antibodies from healthy Caribbean population indicates that 10% of the normal Jamaican population is HTLV positive (36). HTLV antibodies have also been detected (2-10%) in South Africa, Nigeria, Egypt, Tunisia and Ghana (39). Family members of ATLL patients usually have a four-fold higher incidence of HTLV antibodies than the incidence in healthy residents in the immediate vicinity (38, 42). Examination of tissues from subhuman primates has resulted in the identification of HTLV antibodies in different species of monkeys (43). HTLV antibodies have also been detected in Japanese macaques and baboons from an experimental colony in Sukhumi, Russia (39). A retrovirus has been isolated from a cell line established from the peripheral blood of an HTLV seropositive baboon. This monkey virus isolate (PTLV) is closely related to HTLV-1 but can be easily distinguished from HTLV-1 and HTLV-2.

HTLV TRANSMISSION STUDIES

T cells from human umbilical cord blood, bone marrow and adult peripheral blood (16,20,44,45) have been used to infect these cells

Table 3. *Examples of transmission of HTLV-1 and HTLV-2 into human umbilical cord blood T cells* (16,20,59)

Cocultured Cells[a] (Recipient/Donor)	HTLV Expression			
	p19	p24	RT	EM
1. HTLV-1				
C1/MJ	++	++	+	+
C3/MJ	++	++	++	+
C5/MJ	++	++	++	+
C10/MJ	++	++	++	+
C4/UK	++	+++	++	+
C21/MI	++	++	++	+
C91/PL	++	++	++	+
C8/SK	++	+++	+	+
2. HTLV-2				
C344/MO	++	+	++	+
C218/MO	++	+	++	+
C346/MO	++	+	++	+

[a]HTLV transmission into cord blood T cells was carried out by cocultivation of cord blood T cells with X-irradiated (6,000-10,000 R) HTLV positive donor T cells.

with HTLV by either exposing the HTLV positive donor cells with X-rays (6-10,000 R) or treatment with mitomycin-C, and co-cultivating with HTLV negative recipient T cells. After 2-5 weeks in culture, the cultures were examined for the expression of HTLV reverse transcriptase, p19, p24, and type C virus particles by electron microscopy. HTLV belonging to both subgroups 1 and 2 has been transmitted into fresh human cord blood T cells resulting in productive infection and transformation in a majority of the cases (Table 3). These studies show that the HTLV isolates of both subgroups 1 and 2 can infect and replicate in human T cells. Similar studies using cell-free HTLV-1 virus (46) show that HTLV-1 is transmitted at a frequency lower (about 10%) than transmission by co-cultivation. T cells from cord blood or peripheral blood infected with HTLV-1 exhibit an OKT-4 (helper T cell) surface phenotype, whereas the bone marrow cells infected with HTLV-1 (47) expressed both OKT-4 (T helper); OKT-8 (T suppressor), or neither OKT-4 or OKT-8 phenotype, suggesting that various T cell subsets of bone marrow may be involved in HTLV infection.

Nature of HTLV Specific T Cell

The T cells of the patient CR from whose blood cells the first HTLV isolate was obtained were found to be HTLV positive whereas the B cells from this patient were HTLV negative (16,48,49) suggesting that the virus is T cell tropic. Recently, a B cell line has also been obtained from the peripheral blood of an ATLL patient which is positive for both EBV and HTLV. HTLV transmission studies show that in addition to human T cells, T cells from marmosets and rabbits can be infected with HTLV (16,50,51). A comparison of the properties of HTLV infected and uninfected T cells, and HTLV positive T cell lines established from ATLL patients is shown in Table 4. A unique property of the HTLV infected T cells is that they can grow for long periods of time in the presence of low levels or in the absence of exogenously added TCGF. The HTLV infected T cells have TCGF receptors as shown by a monoclonal antibody anti-Tac (52), and are constitutive producers of various lymphokines (53). The surface phenotype and the morphological characteristics, including the presence of convoluted nuclei, of the virus infected T cells are similar to the primary HTLV positive tumor cells (16,49). The HTLV transformed T cells and the fresh tumor cells from the patients do not produce tumors in nude mice (P. Sarin *et al.* unpublished results), and no transformation of NIH-3T3 cells has been observed in the transfection assays using DNA extracted from the

HTLV infected T cells (E. Westin *et al.*, unpublished results).

The T cell lines infected with HTLV release a number of lymphokines
in the conditioned media of these cultures (53). The lymphokines
that have been identified from these T cell lines include: 1) B
cell growth factor (BCGF); 2) colony-stimulating factor (CSF); 3)
differentiation-inducing factor (DIF); 4) eosinophil growth-matura-
tion activity (EGMA); 5) fibroblast activating factor (FAF); 6)
gamma interferon (γ-IFN); 7) interleukin 3 (IL-3); 8) leukocyte-
inhibitory factor (LIF); 9) macrophage activating factor (MAF);
10) migration-enhancement factor (MEF); and 11) macrophage migra-
tion-inhibitory factor (MIF). The HTLV transformed T cell lines do
not release TCGF in the conditioned media although low levels of
TCGF have been detected in some cases (9).

MECHANISM OF TRANSFORMATION BY HTLV

HTLV has the unique property of infecting and transforming T cells
although it is a chronic leukemia virus and does not carry an *onc*
gene. T cells infected with HTLV express high levels of viral mRNA,
including a 9 Kb species probably coding for *gag* and *pol* region, a
4 Kb species that hybridized preferentially to *env* sequences (*env*
mRNA) and a 2 Kb RNA containing the pX and LTR sequences. These
studies point to the possibility that expression of viral proteins,
including the pX region may be essential for transformation.

Another possibility by which HTLV induces leukemogenesis is by the
activation of growth factor genes and c-*onc* gene. Examination of
HTLV positive T cell lines for TCGF expression showed that only
low levels of TCGF mRNA were expressed in a few cases (54), although
high levels of TCGF receptors are observed in HTLV-infected T cells
(52). HTLV infected cells also express a gene HT3 which may be the
gene for the TCGF receptor (Table 4). Examination of the HTLV infected T cells for expression of *onc* genes using cloned probes

Table 4. *Characteristics of HTLV positive and uninfected human T cells* (16,20)

Property	HTLV Positive T Cells		Uninfected Cord Blood T Cells
	Neoplastic	Cord Blood	
1. E-Rosette (T cells)	+++	+++	+++
2. S-IgG[a], EBNA[b], TdT[c]	-	-	-
3. Cell morphology:			
a. Multinucleated gaint cells	++	++	-
b. Lobulated nuclei	++	++	-
4. Cell phenotype: (% positive cells)			
a. Inducer/helper (OKT4)	50-95	70-95	65-95
b. Suppressor/cytotoxic (OKT8)	0-30	0-20	0-40
5. HLA Expression			
a. Additional HLA antigens	+	+	-
b. HLA-DR[d]	+	+	-
6. *In vitro* growth	Indefinite	Indefinite	Limited
7. Need for exogenous TCGF (v/v)	Low to None	Low to None	High
8. TCGF receptor (TAC)[d]	+++	+++	+
9. Expression of p19, p24, RT and type C virus particles (EM)	+	+	-
10. Lymphokine production	+	+	-
11. Expression of HT-3 sequences	+++	+++	+

[a]S-IgG, cell surface immunoglobulins; [b]EBNA, Epstein-Barr virus nuclear antigen; [c]Determined by cell sorter using monoclonal antibodies.

Table 5. *HTLV-3 detection of cultured T cells from patients with AIDS and pre-AIDS* (60,61)

Source of T Cells	# Positive/# Tested	% Positive
1. Pre-AIDS	18/21	86
2. AIDS	26/72	36
a. Children	3/8	38
b. Adults with opportunistic infection	10/21	48
c. Adults with Kaposi sarcoma	13/43	30
3. Healthy homosexuals	1/22	4
4. Healthy heterosexuals	0/115	0
5. Healthy mothers of childhood AIDS cases	3/4	75

HTLV-3 expression was confirmed by the detection of Mg^{++}-dependent reverse transcriptase activity in supernatant fluids; virus particles observed by electron microscopy; and expression of HTLV-3-related antigens by cells as detected with rabbit anti-HTLV-3 or antibody-positive natural sera.

showed that only c-*sis* was expressed in a number of HTLV infected cells. The c-*sis* gene has been shown to code for the gene of the platelet-derived growth factor (PDGF), which normally acts on fibroblasts, smooth muscle cells and glial cells (55,56). Since the autostimulation model appears to be an unlikely mechanism by which HTLV induces transformation, further studies on the molecular mechanism of HTLV transformation will focus on whether the *cis* gene products or other gene products are essential for the initiation and/or maintenance of leukemogenesis. Thus, the HTLV infected T cells provide a unique model system to study the mechanism of transformation by retroviruses.

Table 6.   *Incidence of HTLV-3 antibodies in sera of patients with AIDS, pre-AIDS and other malignancies* (63,64)

Serum Source	# Positive/# Tested	% Positive
1. AIDS patients	77/83	93
2. Pre-AIDS patients	27/34	79
3. IV drug users	4/6	67
4. Homosexuals	8/29	27
a. Sexual contact of AIDS patients	1/8	12
b. Other	6/20	30
5. Normal subjects	1/167	<1
6. Patients with other malignancies:		
a. Hepatitis B infection	0/3	0
b. Rheumatoid arthritis	0/1	0
c. Systematic lupus erythematosis	0/6	0
d. Lymphatic leukemias	3/17	18

ACQUIRED IMMUNE DEFICIENCY SYNDROME (AIDS) AND HTLV

Efforts to isolate an infectious agent involved in AIDS have focused on the possible involvement of retroviruses in the causation of this immunosuppressive disease. Antibodies to various viruses, including cytomegalovirus (CMV), hepatitis B virus (HBV) and Epstein-Barr virus (EBV) have been detected in sera from patients with AIDS (57). Attempts to look for the involvement of HTLV or related retroviruses in AIDS were based on certain similarities between ATLL and AIDS. In ATLL, HTLV is T cell tropic and primarily infects T helper cell population. In AIDS, the depletion of T helper cells is the major defect. Central Africa and Haiti are endemic for both ATLL as

Table 7. *Comparison of the properties of HTLV-3 to HTLV-1 and HTLV-2*

Characteristic	HTLV-1	HTLV-2	HTLV-3
1a. Cell specificity for infection	Lympho-cytes	Lympho-cytes	Lympho-cytes
b. T cell (OKT4$^+$, T helper cell)	+	+	+
2. Presence of giant multi-nucleated cells	+	+	+
3. Major core protein	p24	p24	p24
4. Common p24 epitope	+	+	+
5. Size of reverse transcriptase (RT)	~100K	~100K	~100K
6. RT divalent cation preference	Mg^{++}	Mg^{++}	Mg^{++}
7. Nucleic acid homology to HTLV-1 (stringent conditions)	+++	±	-
8. Nucleic acid homology to HTLV-1 (moderate stringency)	+++	++	+
9. Homology to other retroviruses	-	-	-
10. Presence of pX region	+	+	+

well as AIDS. Close contact is necessary for HTLV transmission as
shown by the presence of HTLV antibodies in the family members of
ATLL patients, and in the case of AIDS, sexual contact is the major
mode of transmission of the AIDS agent. A retrovirus model system
that can cause both leukemia and immune suppression in the host is
the feline leukemia virus. Retroviruses belonging to the HTLV family of retroviruses that have been isolated from AIDS and pre-AIDS
cases include HTLV-1 (58), HTLV-2 (59) and HTLV-3 (60,61) and other
related variants (62). The expression of HTLV-3 in cultured cells
of patients with AIDS and pre-AIDS is summarized in Table 5. Examination of sera from AIDS patients shows that 90% of the patients
are positive for HTLV-3 antibodies (Table 6) as detected by radio-
immunoassays and Western blot analysis (63,64). In the Western blot
analysis, the antibodies in the serum from patients with AIDS and
pre-AIDS appear to be chiefly reactive with a membrane glycoprotein
of 41,000 dalton protein. A comparison of the properties of HTLV-3
with HTLV belonging to subgroups 1 and 2 is shown in Table 7.

The high incidence of HTLV-3 isolation from patients with AIDS and
pre-AIDS, and the establishment of a high producer T cell line
(H_9V) (61) should enable the rapid development of a blood test for
HTLV-3 antibodies for the screening of blood from healthy blood
donors for blood transfusion, and population at risk for development of AIDS. An approach to prevent the transmission and proliferation of HTLV-3 in the infected individuals is to obtain drugs
that will inhibit the virus proliferation. The other approach to
control the disease is to develop a vaccine against this virus
(HTLV-3) for use as a preventive measure for immunization of the
population at risk. Progress in the development of antiviral drugs
for chemotherapy and in the development of a vaccine will be greatly aided by the availability of a suitable animal model system.

REFERENCES

1) WONG-STAAL, F. & GALLO, R.C. (1982). The transforming genes of
   primate and other retroviruses and their human homologs. In:
   "Viral Oncology", G. Klein, ed., Raven Press, New York, pp.
   153-171.

2) JARRETT, W., CRAWFORD, E.M., MARTIN, M.B. & DAVIE, F. (1964).
   Leukemia in the cat. Transmission experiments with leukemia
   (lymphosarcoma), Nature, 202, 567.

3) HARDY, W.D., ZUCKERMAN, A.J., McCLELLAND, H.W., SNYDER, H.W.,
   ESSEX, M. & FRANCIS D, D. (1980). Immunology and eipdemiology
   of feline leukemia virus nonproducer lymphomas. In: "Viruses
   in Naturally Occurring Cancers", M. Essex, G. Todaro, H. zur
   Hausen, eds., Cold Spring Harbor Press, New York, pp. 677-699.

4) BURNY, A., BRUCK, C., CHANTRENNE, H., CLEUTER, Y., DEKEGEL, D.
   GHYSDAEL, J., KETTMANN, R., LECLERG, M., LEUNEN, J., MAMMERICKS,
   M. & PORTETELLE, D. (1980). Bovine leukemia virus: molecular
   biology and epidemiology. In: "Viral Oncology", G. Klein, ed.,
   Raven Press, New York, pp. 231-289.

5) BONNER, R.A., O'CONNEL, C.O. & COHEN, M. (1982). Cloned endogenous retroviral sequences from human DNA. Proc. Natl. Acad.
   Sci. USA, 80, 4709.

6) DUESBERG, P.H. (1979). Transforming genes of retroviruses. Cold
   Spring Harbor Symp. Quant. Biol., 44, 13.

7) HAYWARD, W.S., NEEL, B.G. & ASTRIN, S.M. (1981). Induction of
   lymphoid leukosis by avian leukosis virus: activation of a cellular "onc" gene by promotor insertion. Nature, 290, 475.

8) MORGAN, D.A., RUSCETTI, F.W. & GALLO, R.C. (1976). Selective
   in vitro growth of T-lymphocytes from normal human bone marrow.
   Science, 193, 1007.

9) SARIN, P.S. & GALLO, R.C. (1984). Human T-cell growth factor.
   CRC Crit. Rev. Immunol.,4, 279.

10) POIESZ, B.J., RUSCETTI, F.W., GAZDAR, A.F., BUNN, P.A., MINNA,
    J.A. & GALLO, R.C. (1980). Detection and isolation of type C

retrovirus particles from fresh and cultured lymphocytes of a
patient with cutaneous T-cell lymphoma. Proc. Natl. Acad. Sci.
USA, 77, 7415.

11) POIESZ, B.J., RUSCETTI, F.W., REITZ, M.S., KALYANARAMAN, V.S.
& GALLO, R.C. (1980). Isolation of a new type-C retrovirus
(HTLV) in primary uncultured cells of a patient with Sezary
T-cell leukemia. Nature, 294, 268.

12) TAKATSUKI, K., UCHIYAMA, T., UESHIMA, Y. & HATTORI, T. (1979).
Adult T-cell leukemia: further clinical observations and cyto-
genetic and functional studies of leukemic cells. Jap. J.
Clin. Oncol., 9, 317.

13) UCHIYAMA, T., YODOI, J., SAGAWA, K., TAKATSUKI, K., & UCHINO,
H. (1977). Adult T-cell leukemia: clinical and hematologic
features of 16 cases. Blood, 50, 481.

14) CATOVSKY, D., GREAVES, M.F., ROSE, M., GALTON, D.A.G., GOOLDEN,
A.W.G., McCLUSKEY, D.R., WHITE, J.M., LAMPERT, I., BOURIKAS,
G., IRELAND, R., BROWNELL, A.I., BRIDGES, J.M., BLATTNER, W.A.
& GALLO, R.C. (1982). Adult T-cell lymphoma-leukemia in blacks
from the West Indies. Lancet, i, 639.

15) RHO, H.M., POIESZ, B., RUSCETTI, F.W. & GALLO, R.C. (1981).
Characterization of the reverse transcriptase from a new retrovirus (HTLV) produced by a human cutaneous T-cell lymphoma
cell line. Virology, 112, 355.

16) SARIN, P.S. & GALLO, R.C. (1983). Human T-cell leukemia-lymphoma virus (HTLV). In: "Progress in Hematology", E.B. Brown,
ed., Grune and Stratton, New York, pp. 149-161.

17) REITZ, M.S., POIESZ, B.J., RUSCETTI, F.W. & GALLO, R.C. (1981).
Characterization and distribution of nucleic acid sequences
of a novel type-C retrovirus isolated from neoplastic human
T-lymphocytes. Proc. Natl. Acad. Sci. USA, 78, 1887.

18) KALYANARAMAN, V.S., SARNGADHARAN, M.G., POIESZ, B.J., RUSCETTI,
F.W. & GALLO, R.C. (1981). Immunological properties of a type-
C retrovirus isolated from cultured human T-lymphoma cells and

comparison to other mammalian retroviruses. J. Virol.,81, 906.

19) ROBERT-GUROFF, M., RUSCETTI, F.W., POSNER, L.E., POIESZ, B.J.
& GALLO, R.C. (1981). Detection of the human T cell lymphoma
virus p19 in cells of some patients with cutaneous T cell
lymphoma and leukemia using a monoclonal antibody. J. Exp.
Med., 154, 1957.

20) POPOVIC, M., SARIN, P.S., ROBERT-GUROFF, M., KALYANARAMAN, V.
S., MANN, D., MINOWADA, J. & GALLO, R.C. (1983). Isolation
and transmission of human retrovirus (human T-cell leukemia
virus). Science, 219, 856.

21) MIYOSHI, I., KUBONISHI, I., YOSHIMOTO, S.,AKAGI, T., OHTSUKI,
Y., SHIRAISHI, Y., NAGATA, K. & HINUMA, Y. (1981). Type C
virus particles in a cord T-cell line derived by co-cultivating normal human cord leukocytes and human leukemic T-cells.
Nature, 294, 770.

22) YOSHIDA, M., MIYOSHI, I. & HINUMA, Y. (1982). Isolation and
characterization of retrovirus from cell lines of human adult
T-cell leukemia and its implication in the disease. Proc.
Natl. Acad. Sci. USA, 79, 2031.

23) HAYNES, B.F., MILLER, S.E., MOORE, T.O., DUNN, P.H., BOLOGNESI,
D.P. & METZGAR, R.S. (1983). Identification of human T-cell
leukemia virus in a Japanese patient with adult cell leukemia
and cutaneous lymphomatous vasculities. Proc. Natl. Acad. Sci.
USA, 80, 2054.

24) VYTH-DREES, F.A. & VRIES, J.E. (1983). Human T-cell leukemia
virus in lymphocytes from a T-cell leukaemia patient originating from Surinam. Lancet, ii, 993.

25) GREAVES, M.F., VERBI, W., TILLERY, R., LISTER, T.A., HABESHAW,
J., GUO, H.G., TRAINOR, C.D., ROBERT-GUROFF, M., BLATTNER, W.,
REITZ, M. & GALLO, R.C. (1984). Human T cell leukemia virus
(HTLV) in the United Kingdom. Int. J. Cancer, 33, 795.

26) SARIN, P.S., AOKI, T., SHIBATA, A., OHNISHI, Y., AOYAGI, Y.,
MIYAKOSHI, H., EMURA, I., KAYANARAMAN, V.S., ROBERT-GUROFF, M.,
POPOVIC, M., SARNGADHARAN, M., NOWELL, P.C. & GALLO, R.C.

(1983). High incidence of human type-C retrovirus (HTLV) in family members of a HTLV-positive Japanese T-cell leukemia patient. Proc. Natl. Acad. Sci. USA, 80, 2370.

27) KALYANARAMAN, V.S., SARNGADHARAN, M.G., ROBERT-GUROFF, M., MIYOSHI, I., BLAYNEY, D., GOLDE, D. & GALLO, R.C. (1982). A new subtype of human T-cell leukemia virus (HTLV-II) associated with a T-cell variant of hairy cell leukemia. Science, 218, 571.

28) GOLDE, D.W., QUAN, S. & CLINE, M.J. (1978). Human T-lymphocyte cell line producing colony stimulating factor. Blood, 52, 1068.

29) HAHN, B.H., POPOVIC, M., KALYANARAMAN, V.S., SHAW, G.M., LOMONICO, A., WEISS, S.H., WONG-STAAL, F. & GALLO, R.C. (1984) Detection and characterization of an HTLV-II provirus in a patient with AIDS. In: "UCLA Symposia on Molecular and Cellular Biology". Vol. 16, M.S. Gottlieb & J.E. Groopman, eds., Alan R. Liss, New York, in press.

30) REITZ, M.S., Jr., POPOVIC, M., HAYNES, B.F., CLARK, S.C. & GALLO, R.C. (1983). Relatedness by nucleic acid hybridization of new isolates of human T-cell leukemia-lymphoma virus (HTLV) and demonstration of provirus in uncultured leukemic blood cells. Virology, 126, 688.

31) NAGY, K., CHEINGSONG-POPOV, R. & WEISS, R.A. (1983). Human T cell leukemia virus type 1: Induction of syncytia and inhibition by patient's sera. Int. J. Cancer, 32, 321.

32) NAGY, K., WEISS, R.A., CLAPHAM, P. & CHEINGSONG-POPOV, R. (1984). Human T cell leukemia/lymphoma virus envelope antigens. In: "Human T Cell Leukemia/Lymphoma", R.C. Gallo, M.E. Essex & L. Gross, eds., Cold Spring Harbor Laboratory, New York, pp. 121-130.

33) POPOVIC, M., REITZ, M.S., Jr., SARNGADHARAN, M.G., ROBERT-GUROFF, M., KALYANARAMAN, V.S., NAKAO, Y., MIYOSHI, I., MINOWADA, J., YOSHIDA, M., ITO, Y. & GALLO, R.C. (1982). The virus of Japanese adult T-cell leukaemia is a member of the

human T-cell leukaemia virus group. Nature, 300, 63.

34) OROSZLAN, S., SARNGADHARAN, M.G., COPELAND, T.D., KALYANARAMAN,
V.S., GILDEN, R.V. & GALLO, R.C. (1982). Primary structure
analysis of the major internal protein p24 of human type C T-
cell leukemia virus. Proc. Natl. Acad. Sci. USA, 79, 1291.

35) COPELAND, T.D., OROSZLAN, S., KAYANARAMAN, V.S., SARNGADHARAN,
M.G. & GALLO, R.C. (1983). Complete amino acid sequence of
human T-cell leukemia virus structural protein p15. FEBS, 162,
390.

36) ROBERT-GUROFF, M., SCHUPBACH, J., BLAYNEY, D.W., KALYANARAMAN,
V.S., MERINO, F., LANIER, A., SARNGADHARAN, M.G., CLARK, J.,
SAXINGER, W.C., BLATTNER, W.A. & GALLO, R.C. (1984). Sero-
epidemiologic studies on HTLV-I. In: "Human T-Cell Leukemia-
Lymphoma Viruses", R.C. Gallo, M. Essex & L. Gross, eds.,
Cold Spring Harbor Press, New York, pp. 285-295.

37) BLATTNER, W.A., KALYANARAMAN, V.S., ROBERT-GUROFF, M., LISTER,
T.A., GALTON, D.A.G., SARIN, P.S., CRAWFORD, M.H., CATOVSKY,
D., GREAVES, M. & GALLO, R.C. (1982). The human type-C retro-
virus from the Caribbean region, and relationship to adult
T-cell leukemia/lymphoma. Int. J. Cancer, 30, 257.

38) GALLO, R.C., KALYANARAMAN, V.S., SARNGADHARAN, M.G., SLISKI,
A., VONDERHEID, E.C., MAEDA, M., NAKAO, Y., YAMADA, K., ITO,
Y., GUTENSOHN, N., MURPHY, S., BUNN, P.A., Jr., CATOVSKY, D.,
GREAVES, M.F., BLAYNEY, D.W., BLATTNER, W., JARRETT, W.F.H.,
JEGASOTHY, B.V., JAFFE, E., COSSMAN, J., BRODER, S., FISHER,
R.I., GOLDE, D.W. & ROBERT-GUROFF, M. (1983). Association of
the human type C retrovirus with a subset of adult T cell
cancers. Cancer Res., 43, 3892.

39) SAXINGER, W.C., LANGE-WANTZIN, G., THOMSEN, K., LAPIN, B.,
YAKOVELVA, L., LI, Y.W., GUO, H.G., ROBERT-GUROFF, M.,
BLATTNER, W.A., ITO, Y. & GALLO. R.C. (1984). Human T cell
leukemia virus: A diverse family of related exogenous retroviruses of humans and Old World primates. In: "Human T Cell

Leukemia-Lymphoma Viruses", R.C. Gallo, M. Essex & L. Gross, eds., Cold Spring Harbor Press, New York, pp. 323-330.

40) ROBERT-GUROFF, M., NAKAO, Y., NOTAKE, K., ITO, Y., SLISKI, A. & GALLO, R.C. (1982). Natural antibodies to human retrovirus HTLV in a cluster of Japanese patients with adult T cell leukemia. Science, 212, 975.

41) HINUMA, Y., KOMODA, H., CHOSA, T., KONDO, T., KOHAKURA, M., TAKENAKA, T., KIKSECHI, M., ICHIMARU, M., YUNOKI, K., SATO, I., MATSUO, R., TAKIUCHI, Y., UCHINO, H. & HANAOKA (1982). Antibodies to adult T-cell leukemia-virus-associated antigen (ATLA) in sera from patients with ATL and controls in Japan: a nationwide seroepidemiologic study. Int. J. Cancer, 29, 631.

42) ROBERT-GUROFF, M., KALYANARAMAN, V.S., BLATTNER, W.A., POPOVIC, M., SARNGADHARAN, M.G., MAEDA, M., BLAYNEY, D., CATOVSKY, D., BUNN, P.A., SHIBATA, A., NAKAO, Y., ITO, Y., AOKI, T. & GALLO, R.C. (1983). Evidence for human T cell lymphoma-leukemia virus infection of family members of human T cell lymphoma-leukemia virus positive T cell leukemia-lymphoma patients. J. Exp. Med., 157, 248.

43) MIYOSHI, I., YOSHIMOTO, S., FUJISHITA, M., HIROKUNI, T., KUBONISHI, I., NIIYA, K. & MINEZAWA, M. (1982). Natural adult T-cell leukemia virus infection in Japanese monkeys. Lancet, ii, 658.

44) POPOVIC, M., LANGE-WANTZIN, G., SARIN, P.S., MANN, D.& GALLO, R.C. (1983). Transformation of human umbilical cord blood T cells by human T-cell leukemia/lymphoma virus. Proc. Natl. Acad. Sci. USA, 80, 5402.

45) MARKHAM, P.D., SALAHUDDIN, S.Z., KALYANARAMAN, V.S., POPOVIC, M., SARIN, P. & GALLO, R.C. (1983). Infection and transformation of fresh human umbilical cord blood cells by multiple sources of human T-cell leukemia-lymphoma virus (HTLV). Int. J. Cancer, 31, 413.

46) RUSCETTI, F.W., ROBERT-GUROFF, M., CECCHERINI-NELLI, L., MINODA, J., POPOVIC, M. & GALLO, R.C. (1983). Persistent in

vitro infection by human T-cell leukemia-lymphoma virus (HTLV) or normal human T-lymphocytes from blood relatives of patients with HTLV associated mature T-cell neoplasms. Int. J. Cancer, 31, 171.

47) MARKHAM, P.D., SALAHUDDIN, S.Z., MACCHI, B., ROBERT-GUROFF, M. & GALLO, R.C. (1984). Transformation of different phenotypic types of human bone marrow T-lymphocytes by HTLV-I. Int. J. Cancer, 33, 13.

48) GALLO, R.C., MANN, D., BRODER, S., RUSCETTI, F.W., MAEDA, M., KALYANARAMAN, V.S., ROBERT-GUROFF, M. & REITZ, M.S., Jr., (1982). Human T-cell leukemia-lymphoma virus (HTLV) is in T but not B lymphocytes from a patient with cutaneous T-cell lymphoma. Proc. Natl. Acad. Sci. USA, 79, 5680.

49) GALLO, R.C., WONG-STAAL, F. & SARIN, P.S. (1984). Cellular *onc* genes, T-cell leukaemia-lymphoma virus, and leukaemias and lymphomas of man. In: "Mechanisms of Viral Leukaemogenesis", J.M. Goldman & J.O. Jarrett, eds., Churchill Livingstone, London, pp. 11-37.

50) MIYOSHI, I., TAGUCHI, H., FUJISHITA, M., YOSHIMOTO, S., KUBONISHI, I., OHTSUKI, Y., SHIRAISHI, Y. & AKAGI, T. (1982). Transformation of monkey lymphocytes with adult T-cell leukemia virus. Lancet, i, 1016.

51) MIYOSHI, I. YOSHIMOTO, S., TAGUCHI, H., KUBONISHI, I., FUJISHITA, M., OHTSUKI, Y. & SHIRAISHI, Y. (1983). Transformation of rabbit lymphocytes with T-cell leukemia virus. Gann, 74, 1.

52) WALDMANN, T., BRODER, S., GREENE, W., SARIN, P.S., SAXINGER, C., BLAYNEY, D., BLATTNER, W., GOLDMAN, C., FROST, K., SHARROW, S., DEPPER, J., LEONARD, W., UCHIYAMA, T. & GALLO, R.C. (1984). A functional and phenotypic comparison of human T cell leukemia/lymphoma virus (HTLV) positive adult T cell leukemia with HTLV negative Sezary leukemia and their distinction using anti-TAC, a monoclonal antibody identifying the human receptor

for T cell growth factor. <u>J. Clin. Invest.</u>,<u>73</u>, 1711.

53) SALAHUDDIN, S.W., MARKHAM, P.D., LINDNER, S.G., GOOTENBERG, J., POPOVIC, M., HEMMI, H., SARIN, P.S. & GALLO, R.C. (1984). Lymphokine production by cultured human T-cells transformed by human T-cell leukemia-lymphoma virus-I. <u>Science</u>, <u>223</u>,703.

54) CLAR, S.C., ARYA, S.K., WONG-STAAL, F., MATSUMOTO-KOBAYASHI, M., RAY, R.M., KAUFMAN, R.J., BROWN, E.L., SHOEMAKER, C., COPELAND, T., OROSZLAN, S., SMITH, K., SARNGADHARAN, M.G., LINDNER, S.G. & GALLO, R.C. (1984). Human T-cell growth factor: partial amino acid sequence, cDNA cloning, and organization and expression in normal and leukemic cells. <u>Proc. Natl. Acad. Sci. USA</u>, <u>81</u>, 2543.

55) DOOLITTLE, R.F., HUNKAPILLAR, M.W., HOOD, L.E., DEVARE, S.G., ROBBINS, K.C., AARONSON, S.A. & ANTONIADES, H.N. (1983). Simian sarcoma virus *onc* gene, v-*sis*, is derived from the gene (or genes) encoding a platelet-derived growth factor. <u>Science</u>, <u>221</u>, 275.

56) WATERFIELD, M.D., SCRACE, G.T., WHITTLE, N., STROOBANT, P., JOHNSSON, A., WASTESON, A., WESTERMARK, B., HELDIN, C.H., HUANG, J.S. & DEUL, T.F. (1983). Platelet derived growth factor is structurally related to the putative transforming protein p24 of simian sarcoma virus. <u>Nature</u>, <u>304</u>, 35.

57) SARIN, P.S. & GALLO, R.C. (1984). Role of viruses in the etiology of acquired immune deficiency syndrome. In: "A Basic Guide for Clinicians", P. Ebbesen, R. Biggar & M. Melbye, eds., Munksgaard, Copenhagen, pp. 173-183.

58) GALLO, R.C., SARIN, P.S., GELMANN, E.P., ROBERT-GUROFF, M., RICHARDSON, E., KALYANARAMAN, V.S., MANN, D. SIDHU, G.D., STAHL, R.E., ZOLLA-PAZNER, S., LEIBOWITCH, J. & POPOVIC, M. (1983). Isolation of human T-cell leukemia virus in acquired immune deficiency syndrome (AIDS). <u>Science</u>, <u>220</u>, 865.

59) POPOVIC, M., KALYANARAMAN, V.S., MANN, D.L., RICHARDSON, E., SARIN, P.S. & GALLO, R.C. (1984). Infection and transformation

of T cells by human T cell leukemia/lymphoma virus of sub-
groups I and II (HTLV-I, HTLV-II). In: "Human T-Cell Leukemia-
Lymphoma Viruses", R.C. Gallo, M. Essex & L. Gross, eds.,
Cold Spring Harbor Press, New York, pp. 217-227.

60) GALLO, R.C., SALAHUDDIN, S.Z., POPOVIC, M., SHEARER, G.M.,
KAPLAN, M., HAYNES, B.F., PALKER, T.J., REDFIELD, R., OLESKE,
J., SAFAI, B., WHITE, G., FOSTER, P. & MARKHAM, P.D. (1984).
Frequent detection and isolation of cytopathic retroviruses
(HTLV-III) from patients with AIDS and at risk for AIDS.
Science, 224, 500.

61) POPOVIC, M., SARNGADHARAN, M.G., READ, E. & GALLO, R.C. (1984).
Detection, isolation, and continuous production of cytopathic
retroviruses (HTLV-III) from patients with AIDS and pre-AIDS.
Science, 224, 500.

62) BARRE-SINOUSSI, F., CHERMANN, J.C.,REY, F., NUGEYRE, M.T.,
CHAMARET, S., GRUEST, J., DAUGET, C., AXLER-BLIM, C., VEZINET-
BRUN, F., ROUZIOUS, C., ROZENBAUM, W. & MONTAGNIER, L. (1983).
Isolation of a T-lymphotropic retrovirus from a patient at
risk for AIDS. Science, 220, 868.

63) SARNGADHARAN, M.G., POPOVIC, M., BRUCH, L., SCHUPBACH, J. &
GALLO, R.C. (1984). Antibodies reactive with human T-lympho-
tropic retroviruses (HTLV-III) in the serum of patients with
AIDS. Science, 224, 506.

64) SCHUPBACH, J., POPOVIC, M., GILDEN, R.V., GONDA, M.A.,
SARNGADHARAN, M.G. & GALLO, R.C. (1984). Serological analysis
of a subgroup of human T-lymphotropic retroviruses (HTLV-III)
associated with AIDS. Science, 224, 503.

CONSTRUCTION OF PROTEIN DATABASES FOR COMPARISON OF NORMAL AND TRANSFORMED CELLS

James I. Garrels & B. Robert Franza, Jr.

Cold Spring Harbor Laboratory
Cold Spring Harbor, N.Y. 11724

INTRODUCTION

Transformation of cultured cell lines and primary cells can be achieved by defined agents including cloned viral genes and activated cellular oncogenes. The transformed cell is recognized by its altered morphology and relaxed growth regulation in culture, and not necessarily by its tumorigenicity. Relationships between some of the oncogene products and biochemical pathways of growth regulation are just beginning to emerge.

Although the protein products of some of the transforming genes, such as SV40 T-antigen (1) and activated cellular p21-ras (2,3), have been identified and partially characterized, the biochemical pathways of growth control and the proteins involved remain largely unknown. Of the many proteins that are normally induced or repressed as cells pass from a proliferative to a quiescent state, some might fail to be induced or repressed in transformed cells. An examination of regulatory patterns for many proteins might be more revealing than simply comparing protein levels in normal and transformed cells at one state of growth.

Two-dimensional gel electrophoresis has been useful for the detection and comparison of large numbers of proteins among cultured cell lines (4-7). Several interesting new proteins have been
found by this approach, including the pp36 protein (8) and the
PCNA/cyclin protein (9,10). Previously unsuspected changes of major cytoskeletal proteins have been detected (11,12) and some of
the oncogene products have been identified (12). A more complete
analysis of cellular proteins by two-dimensional gel electrophoresis has been hampered by the difficulties in quantitation, pattern
comparison, and data handling.

THE PROTEIN DATABASE APPROACH

We have worked intensively on the problems of standardized two-
dimensional gel electrophoresis and computer analysis (13,14).
Working with the rat REF52 cell system (15,16), we have endeavor-
ed to build a database of protein information for normal and transformed cells. Such a database consists of quantitative two-dimensional gel data from many experiments, all referenced to a standardized 2D gel pattern. The standard spot numbers can be used
to record protein identifications and accessory information such
as subcellular localization and post-translational modification.
Most importantly, the quantitative data can be used to plot the
regulatory behavior of any detected protein during the course of
multiple experiments. Here we use the database approach to begin
to elucidate, in normal and transformed REF52 cells, some of the
changes that occur in each line as cells reach confluent densities in cell culture.

A DATABASE FOR A NORMAL AND A TRANSFORMED CELL LINE

In this experiment, REF52 cells and AG6 cells, an SV40-transform-
ed derivative of REF52 cells, were analyzed under identical culture conditions at several different times after plating. Expo-

nentially dividing cells were labeled one day after plating, late log phase cells were labeled 4 days after plating, and completely confluent cells were labeled 8 days after plating. At day 8, normal cells had become completely quiescent, and the transformed cells were dividing very slowly. At each time point, cells were labeled for 24 hr with approximately 250 uCi/ml of [^{35}S]-methionine.

Two-dimensional gels of each sample were run, using several different combinations of pH gradient and acrylamide concentration. Shown in Figure 1 are portions of pH 3.5-10, 10% acrylamide gels showing proteins synthesized in dividing and confluent cells for each line. Over 1500 proteins were resolved on these gels.

These two-dimensional gels, and their calibration gels, were scanned and processed by the QUEST software for spot detection and integration (14). A reconstructed (synthetic) image (Figure 2) is produced by the computer to show the relative position and size of each detected spot. Note that the spots are accurately detected, and most overlapping clusters are correctly resolved. For spots that have completely blackened the film, data is taken from the scan of a shorter exposure. For every spot detected by the computer, complete quantitative data is recorded including position, integrated density, integrated disintegrations per minute (dpm), Gaussian fitting parameters, degree of overlap with neighboring spots, and more. The integrated dpm is indicated beside each spot in Figure 2.

To plot the intensity of any protein throughout the course of an experiment, the spot patterns must be matched. Automatic matching programs begin the process by finding landmark spots, and these in turn are used as reference points for the complete matching process. Interactive (manual) matching programs are used only for visual confirmation and minor editing if necessary.

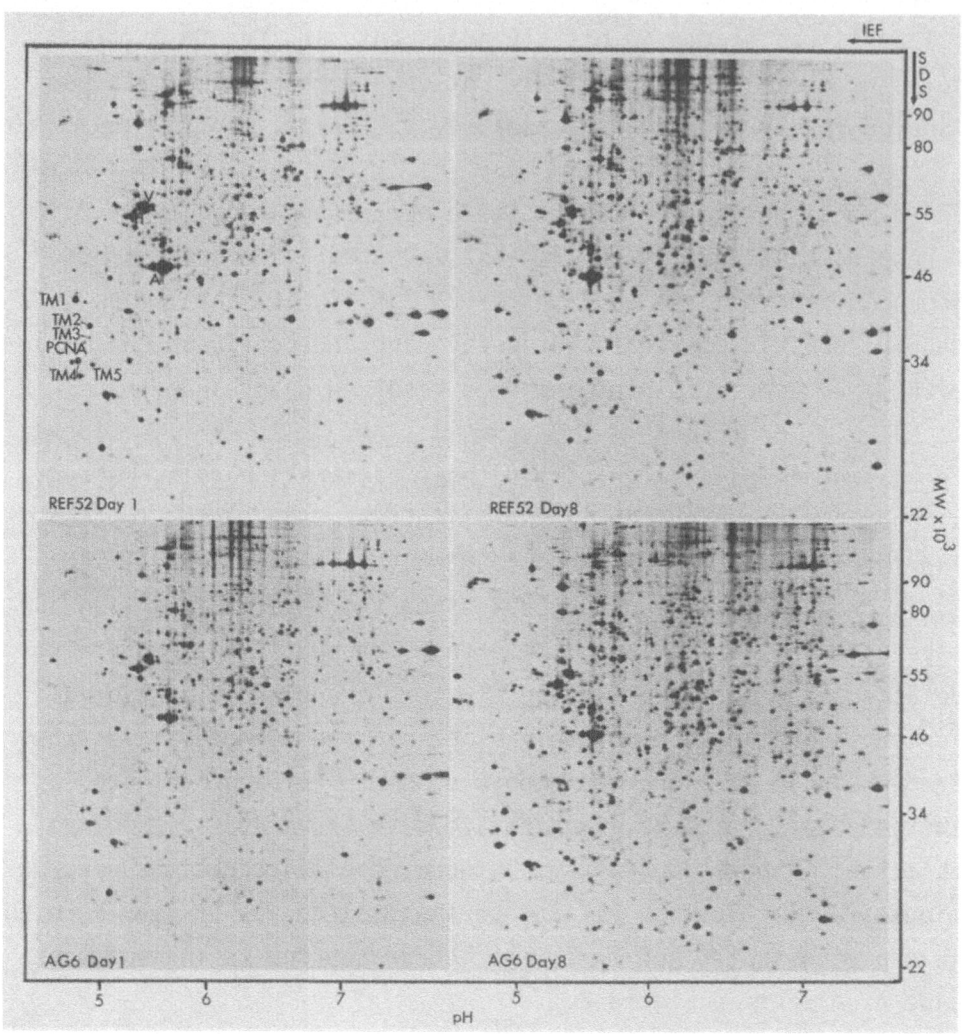

Fig. 1. *Two-dimensional gel patterns of REF52 and AG6 cells during proliferative growth (day 1) and at confluence (day 8).* The indicated protein are A, actin; V, vimentin; T, tubulin; TM1-5, tropomyosins; and PCNA, proliferating cell nuclear antigen or cyclin.

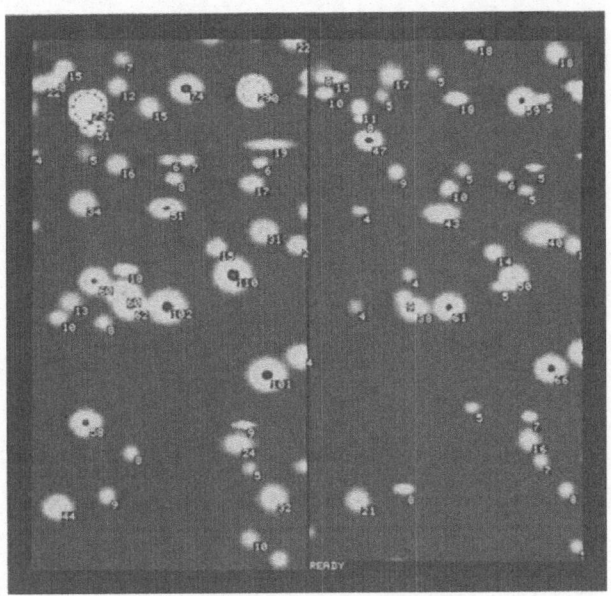

Fig. 2. *Computer-generated image of spots detected and quantitated from REF52 cells (left) and AG6 cells (right) at day 1.* The region shown is slightly below and to the right of actin. The numbers indicated beside each spot are the total dpm detected in each spot.

To match the entire group of gels run as part of one experiment, one of the gels is arbitrarily selected as the standard pattern and each of the other patterns is matched to it. Spots not present in the original standard are added to it as matching progresses. In the end, the standard image represents all spots detected in the experiment. An example of the match of a film to the standard is shown in Figure 3.

When the proteins from each sample in an experiment have been quantified and matched to a standard pattern, a database has been formed. Each protein in the database can be addressed by sample number and by a 4-digit standard spot number. The quantitative

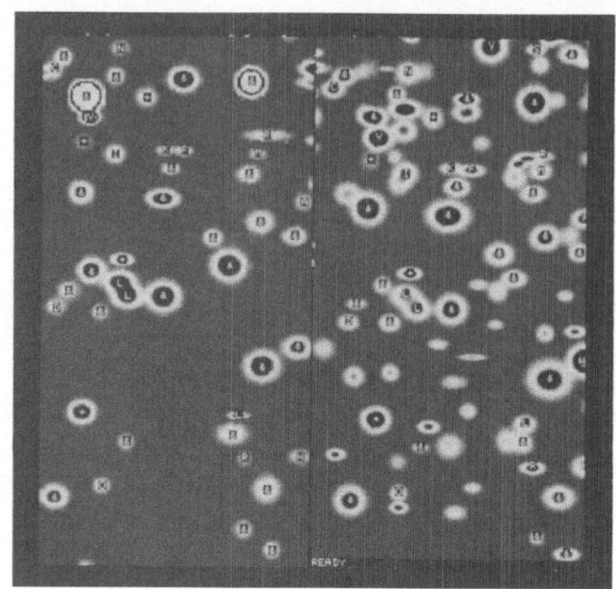

Fig. 3. *Matching of a spot pattern to the standard pattern*. Shown is a portion of the match between the protein pattern of REF52 day 1 (left) and the standard pattern (right). All spots were matched automatically; those labeled '&' served as the landmarks from which the remaining pattern was matched.

value stored for each spot is the disintegrations per minute (dpm) detected in a spot divided by the dpm loaded on that particular gel. This normalization compensates for unequal loading of gels and converts the detected intensity of each protein to a biologically relevant quantity. If labeling times were short, the value recorded for each protein represents the fraction of total protein synthesis; if labeling times were long, the value recorded represents steady-state abundance of the protein. A database derived from one experiment can be further extended by matching other experiments to the same standard, or by matching the standard pattern to other standard patterns used for other experiments. Once data has been entered into a database, it can be searched, tabulated, and graphed in several ways, as seen below.

Table 1. *Relatedness of normal and transformed cells during proliferating and at confluence*

Comparison	Standard deviation(a)	Differences > 2-fold(b)	Qualitative differences(c)
REF52 dividing vs AG6 dividing	0.97	18%	9
REF52 dividing vs REF52 confluent	1.11	22%	4
AG6 dividing vs AG6 confluent	1.01	22%	7
REF52 confluent vs AG6 confluent	1.26	29%	13

(a) Based on logarithm (base 2) of spot ratios (approx. 650 spots compared).

(b) Percent of matched spots with ratios less than 0.5 or greater than 2.0 (approx. 650 spots compared).

(c) Proteins prominent in one sample (approx. 300 most abundant) but not detected in other sample.

EXTRACTION OF RESULTS FROM THE DATABASE

The first question often asked of an experiment is "How related are the samples to one another?" How much did the protein patterns change due to density, how much due to transformation, and are there any new proteins? In Table 1, several measures of related-ness are presented to compare 1) dividing REF52 vs dividing AG6 cells, 2) dividing REF52 vs confluent REF52 cells, 3) dividing AG6 vs confluent AG6 cells, and 4) confluent REF52 vs confluent AG6 cells.

A statistical measure of relatedness, the standard deviation of
the logarithm (base 2) of spot ratios, takes into account both
the number and the magnitude of the differences. Duplicate samples
give a standard deviation value of approximately 0.5. A second
measure of relatedness is a simple count of the proteins that
differ by more than a factor of 2. Each of the above comparisons
was made on the basis of approximately 650 matched pairs of spots.
The final comparison is a count of the qualitative differences.
These are the spots present in one gel but undetected in the other.
To be considered, however, a spot must be seen substantially above
the minimum level of detection. Here only spots with intensity 10-
fold greater than the sensitivity were considered; there were
about 300 such spots in each gel.

The data in Table 1 reveals that the differences due to growth
state (dividing vs confluent in either line) are greater than the
differences between proliferating cultures of REF52 and AG6. However, at confluence, the two lines become quite divergent differing quantitatively and qualitatively by much more than they did
at low density. This data shows that both lines change as they
reach confluence, but that the changes for each line are substantially different.

To find the individual proteins that differ between normal and
transformed cells or between lines at different density, we can
ask the computer to create and display spot sets. One set was
created to contain the proteins that increase at confluence in
normal cells (Figure 4, left panel) and another set was created
to contain the proteins that increase at confluence in transformed cells (Figure 4, right panel). On the color screen, members of a set are highlighted in color; on the photograph they
appear in solid black. It can be seen, even from the small region
of the image shown, that the two sets are substantially different.

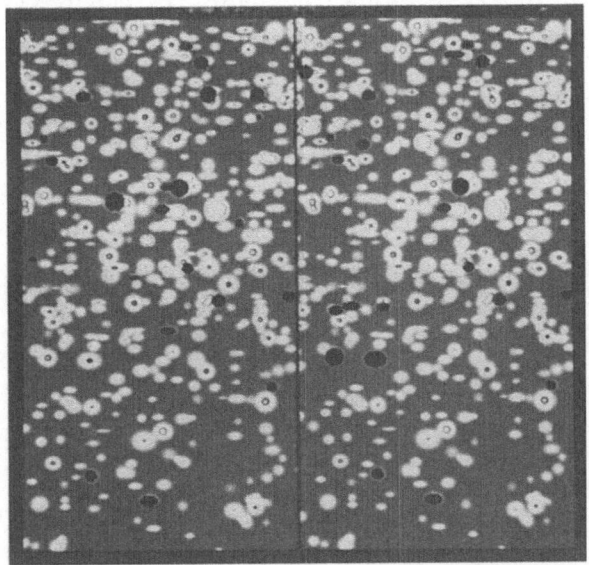

Fig. 4. *Spot sets indicating proteins that increase at confluence.*
Proteins that increase by 2-fold or more in REF52 cells (left) or
in AG6 cells (right) are indicated as filled black areas on the
standard protein pattern. The sets, generated from the database,
reveal common changes and changes unique to each cell line.

Any spot of interest can be further examined graphically. In Figure 5, data for selected spots is shown graphically at the top of
the screen. Points A, B, and C for each graph show REF52 cells at
days 1, 4, and 8, respectively; points D, E, and F show AG6 cells
at the same days after plating. One spot (spot number 1330) increases in both lines at confluence, although the baseline levels
are different. Two of the spots shown (spot numbers 2360 and 8350)
are highly regulated in REF52 cells, but regulated very little
during growth in AG6 cells. Two very interesting spots (spot numbers 3390 and 3348) are barely detectable and apparently unregu-

lated by growth in REF52 cells; however, in AG6 cells they are
dramatically induced at confluence.

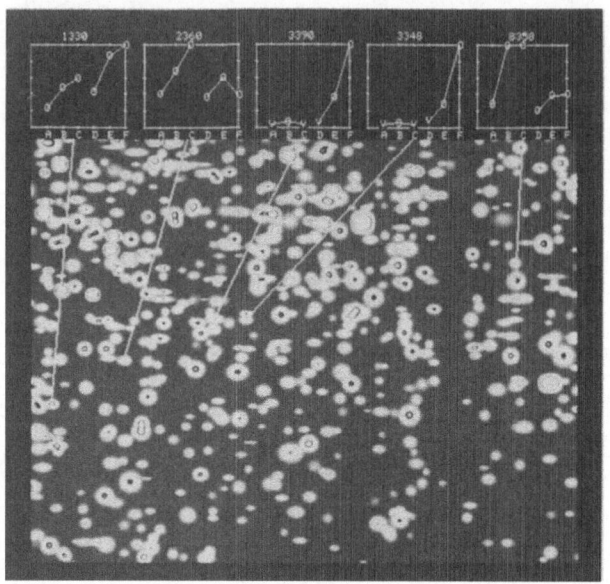

Fig. 5. *Graphical analysis of selected spots.* Data from the database is plotted for five selected spots (standard spot numbers
above each graph). Each graph shows only relative change throughout the course of the experiment, although the absolute intensity
of each spot in each sample can be obtained.

Much more information is available from the analysis of these
samples than can be presented here. For example, changes in the
group of tropomyosin proteins and PCNA are apparent from Figure 1.
Additional results from this and other related experiments will
be presented elsewhere.

DISCUSSION

One objective of this comparison of normal and SV40-transformed
REF52 cells was to find proteins that respond differently in the
two lines as confluent densities are reached. As expected, some
proteins were found that are growth-regulated in normal but not in
transformed cells. Somewhat surprisingly, numerous proteins were
also found which are growth-regulated in transformed but not in
normal cells. The overall changes at confluence are approximately
equal, but in terms of individual proteins the responses to confluence are quite divergent. During proliferative growth, REF52
and AG6 cells show highly related protein patterns, but at confluence, the patterns are much less related.

The analysis presented here is preliminary and is intended to demonstrate the construction and use of 2D gel databases. Many related
experiments using normal and transformed REF52 cells have been
carried out in our laboratory, and much more detailed databases
are being constructed. The prospect of large and highly informa-
tive protein databases is now made realistic by 1) the development of a complete software system for 2D gel quantitation, pattern
matching, and data management; 2) a large number of gels run using
a standardized 2D gel system; and 3) the falling cost of powerful
computer and graphics equipment.

With a more complete database for rat cells, one could ask additional questions of any protein of potential interest. For example,
for protein 3390 one could ask: "Is it induced at confluence in
cells transformed by other agents?" "Is it induced in normal
cells depleted of nutrients or other serum factors?" "Is it a
phosphorylated derivative of spot 3348?" "Is it a part of the
response to any purified growth factor?" "Is it a nuclear protein?"
"Are any other proteins coregulated with it?".

The database approach should point out many such proteins of
interest and can answer some fundamental questions about their
regulation before intensive biochemical investigations are begun.
Detection and quantitation on two-dimensional gels is no substitute
for protein purification followed by painstaking structural and
functional studies, but the databasis can potentially aid in such
efforts by revealing tissues of highest abundance, by suggesting
preferred fractionation procedures, and by indicating levels of
modification or microheterogeneity. The development of larger databases containing much information about the proteins of normal and
transformed cells should aid considerably in our understanding of
the biochemistry of growth control and cellular transformation.

REFERENCES

1) TJIAN, R. (1981). Regulation of viral transcription and DNA
   replication by the SV40 large T antigen. Curr. Topics Microbiol. Immunol., 93, 5.
2) FURTH, M.E., DAVIS, L.J., FLEURDELYS, B., & SCOLNICK, E.M.
   (1982). Monoclonal antibodies to the p21 products of the
   transforming gene of Harvey Murine Sarcoma Virus and of the
   cellular ras gene family. J. Virol., 43, 294.
3) PAPAGEORGE, A., LOWY, D., & SCOLNICK, E.M. (1982). Comparative
   biochemical properties of p21 ras molecules coded for by viral
   and cellular ras genes. J. Virol., 44, 509.
4) GARRELS, J.I. (1979). Two-dimensional gel electrophoresis and
   computer analysis of proteins synthesized by clonal cell lines.
   J. Biol. Chem., 254, 7961.
5) BRAVO, R. & CELIS, J.E. (1980). Gene expression in normal and
   virally transformed mouse 3T3B and hamster BHK21 cells. Exp.
   Cell Res., 127, 249.

6) LEAVITT, J., GOLDMAN, D., MERRIL, C., & KAKUNAGA, T. (1982). Actin mutations in a human fibroblast model for carcinogenesis. Clin. Chem., 28, 850.

7) FRANSEN, L., VAN ROY, F., & FIERS, W. (1983). Changes in gene expression and protein phosphorylation in murine cells, transformed or abortively infected with wild type and mutant Simian Virus 40. J. Biol. Chem., 258, 5276.

8) RADKE, K. & MARTIN, G.S. (1979). Transformation by Rous sarcoma virus: Effects of src gene expression on the synthesis and phosphorylation of cellular polypeptides. Proc. Natl. Acad. Sci. USA, 76, 5212.

9) BRAVO, R., FEY, S.J., BELLATIN, J., LARSEN, P.M., AREVALO, J., & CELIS, J.E. (1981). Identification of a nuclear and of a cytoplasmic polypeptide whose relative proportions are sensitive to changes in the rate of cell proliferation. Exp. Cell Res., 136, 311.

10) MATHEWS, M.B., BERNSTEIN, R.M., FRANZA, B.R., & GARRELS, J.E. (1984). Identity of the proliferating cell nuclear antigen and cyclin. Nature, 309, 374.

11) MATSUMURA, F., LIN, J.J.-C., YAMASHIRO-MATSUMURA, S., THOMAS, G.P., & TOPP, W.C. (1983). Differential expression of tropomyosin forms in the microfilaments isolated from normal and transformed rat cultured cells. J. Biol. Chem., 258, 13954.

12) FRANZA, B.R. & GARRELS, J.E. (1984). Transformation-sensitive proteins of REF52 cells detected by computer-analyzed two-dimensional gel electrophoresis. Cancer Cells, 1, 133.

13) GARRELS, J.I. (1983). Quantitative two-dimensional gel electrophoresis of proteins. Meth. Enz., 100, 411.

14) GARRELS, J.I., FARRAR, J.T., & BURWELL, C.B. (1984). The QUEST system for computer-analyzed two-dimensional electrophoresis of proteins. In: Two-Dimensional Gel Electrophoresis of Proteins: Methods and Applications (eds J.E. Celis & R. Bravo), p. 37, Academic Press, New York.

15) MCCLURE, D.B., HIGHTOWER, M.J., & TOPP, W.C. (1982). Effect
    of SV40 transformation on the growth factor requirements of
    the rat embryo cell line REF52 in serum-free medium. <u>Cold
    Spring Harbor Conf.</u>, <u>Cell Proliferation</u>, <u>9</u>, 345.

16) McCLURE, D.B., DERMODY, M., & TOPP, W.C. (1984). In vitro
    correlates of tumorigenicity of REF52 cells transformed by
    Simian Virus 40. <u>Cancer Cells</u>, <u>1</u>, 17.

CYCLIN (PCNA) IS A COMPONENT OF THE PATHWAY(S) LEADING TO DNA REPLICATION AND CELL DIVISION: A ROLE IN DNA REPLICATION?

Julio E. Celis & *Ariana Celis*

Division of Biostructural Chemistry, Department of
Chemistry, Aarhus University, DK-8000 Aarhus C, Denmark

INTRODUCTION

Understanding of the molecular mechanisms underlying malignant
transformation and cancer will be assisted by the identification
of cellular proteins whose activity may be involved in the control
of cell proliferation in normal cells (1-3 and references therein). The acidic nuclear polypeptide cyclin (M_r=36,000; IEF 49 in
the HeLa protein catalogue, 4-6; see also Fig. 1) is potentially
such a candidate as the levels of this protein are modulated during the cell cycle (increase in S-phase; 7) and correlate directly with the proliferative state of normal cells. Cyclin is pres-
end in very small amounts in normal non-dividing cells (senescent
and quiescent cells included) and tissues, but is synthesized by
proliferating cells both of normal and transformed origin, tumours included (1,7-25). So far, most of the properties of cyclin
are also shared by the proliferating cell nuclear antigen (PCNA;
26-29), a human protein that has recently been shown to be identical with cyclin (21).

Here we present a detailed immunofluorescence study of the distribution of cyclin during the cell cycle of transformed human

223

Fig. 1. *Two dimensional gel electrophoresis (IEF, NEPGHE) of [^{3}S]-methionine labelled proteins from asynchronous HeLa cells labelled for 24 hr.* In IEF, the pH ranges from 7.5 (left) to 4.5 (right). In NEPHGE, the pH varies from 7.5 (right) to 10.5 (left). 1,357 proteins (946 acidic and 411 basic) are indicated in this Figure. Cyclin corresponds to IEF 49. From Bravo *et al.* (4).

amnion cells (AMA) using PCNA autoantibodies that react specifically with this protein (21). The results confirm and extend previous gel electrophoretic (7) and immunofluorescence studies (27,
30) which showed that cyclin is preferentially synthesized during
S-phase. Furthermore, they reveal dramatic changes in its distribution during the cell cycle, particularly within S-phase.
Some of these patterns (nucleolar exclusion; nucleolar labelling)
are strikingly similar to those observed in cells labelled with
[^{3}H]-thymidine, and thus it seems likely that the distribution of
this protein may reflect at least some of the stages of DNA replication (32-34). Taken together, these results support the notion that cyclin is a central component of the pathway(s) leading
to cell division and that its activity may be associated with specific aspects of DNA replication (1,18,19,30,31,34, and references
therein).

IMMUNOFLUORESCENCE LOCALIZATION OF CYCLIN IN ASYNCHRONOUS AMA
CELLS

Fig. 2b shows an indirect immunofluorescence micrograph of asynchronous transformed human amnion cells (AMA) permeated and fixed
with methanol and incubated with PCNA antibodies that react specifically with cyclin (21). About 42% of the total cell population reacts strongly with the antibody (compare Figs 2a and b)
revealing a variable nuclear staining both in terms of intensity
and distribution of the antigen (Figs 2b and c) (30). Very little
cytoplasmic staining can be observed. Some nuclear patterns of
cyclin staining are indicated with arrows in Fig. 2c, and a description of these patterns as well as their putative sequence of
appearance during the cell cycle is given below (30). Evidence
suggesting that the differential nuclear staining observed with
PCNA antibodies is due to cell cycle variations and not to per-
meation or fixation artifacts has been obtained by double immunofluorescence using PCNA antibodies (Fig. 2c) and a monoclonal

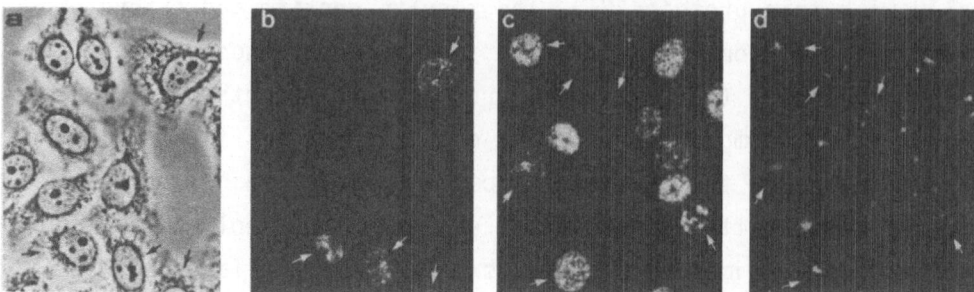

Fig. 2. *Nuclear localization of cyclin in asynchronous AMA cells.*
(a,b), phase contrast (a) and epifluorescence (b; PCNA antibodies)
micrographs of the same field of cells treated with methanol prior
to immunofluorescence. (c,d), double immunofluorescence of
methanol-treated cells reacted with PCNA antibodies (c) and a
mouse monoclonal antibody that reacts with the nucleus of all
interphase cells (d). From Celis and Celis (30).

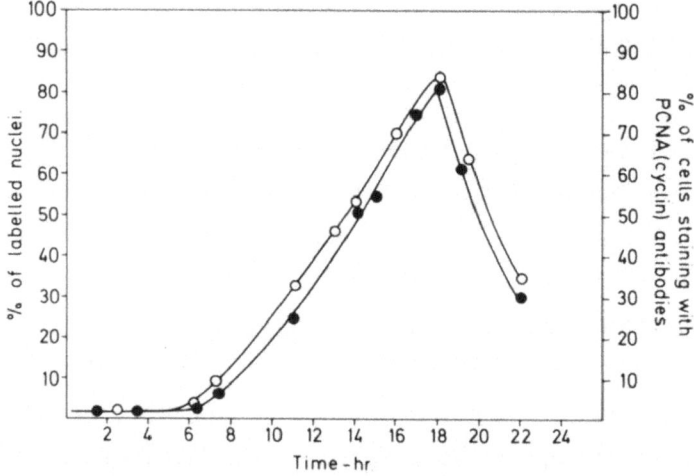

Fig. 3. [^5H]-*thymidine incorporation (●-●) and cyclin staining
(o-o) in synchronous AMA cells.* Cells (grown in coverslips) with-
drawn at various times after plating mitotic cells were labelled
with [^{3}H]-thymidine (30 min.; 2 µCi/ml) and prepared for immunofluorescence as previously described (22,30). From Celis and
Celis (30).

antibody that stains the nucleus, including nucleoli, of all
interphase AMA cells (Fig. 2d).

SEQUENCE OF APPEARANCE OF THE VARIOUS CYCLIN STAINING PATTERNS THROUGH THE CELL CYCLE OF AMA CELLS - SUBDIVISION OF S-PHASE

To determine the sequence of appearance of the different cyclin
staining patterns through the cell cycle, synchronized mitotic
AMA cells obtained by mechanical detachment were plated on glass
coverslips and analyzed at various times for $[^3H]$-thymidine incorporation (autoradiography; •-•, Fig. 3) and indirect immunofluorescence using PCNA antibodies (o-o, Fig. 3) (30). The results, which are presented in Fig. 3, showed a close correlation
between % of cells exhibiting DNA synthesis and positive cyclin
staining, suggesting that only S-phase cells stain strongly with
the antibody (30). Similar observations have been reported by
Takasaki *et al*. (27) in human lymphoid cells.

A putative sequence of cyclin staining patterns deduced from observations of sister AMA cells is presented in Fig. 4 (30). It
should be stressed that the transition between the different
staining patterns does not take place simultaneously in all cells,
and that in most cases, the assignments have been aided by the
observation of slightly asynchronous sister cells as well as of
multinucleated cells showing slightly asynchronous nuclear staining patterns. During G1 (1 to 6½ hr after plating; see also Fig.
3), cyclin staining is weak but characteristic, and it is mainly
confined to defined nuclear structures of unknown origin (Fig.
4a). 7 to 7½ hr after plating (Fig. 3; beginning of S-phase), the
first cells showing increased cyclin staining are observed (Fig.
4b). In these cells, cyclin (granular pattern) is found throughout the nucleus with the exception of the nucleoli (Fig. 4b),
suggesting that it is located mainly within the nucleus and not
confined to the nuclear envelope. The absence of nucleolar stain-

Cell cycle phase   **Patterns of cyclin (PCNA) staining.**

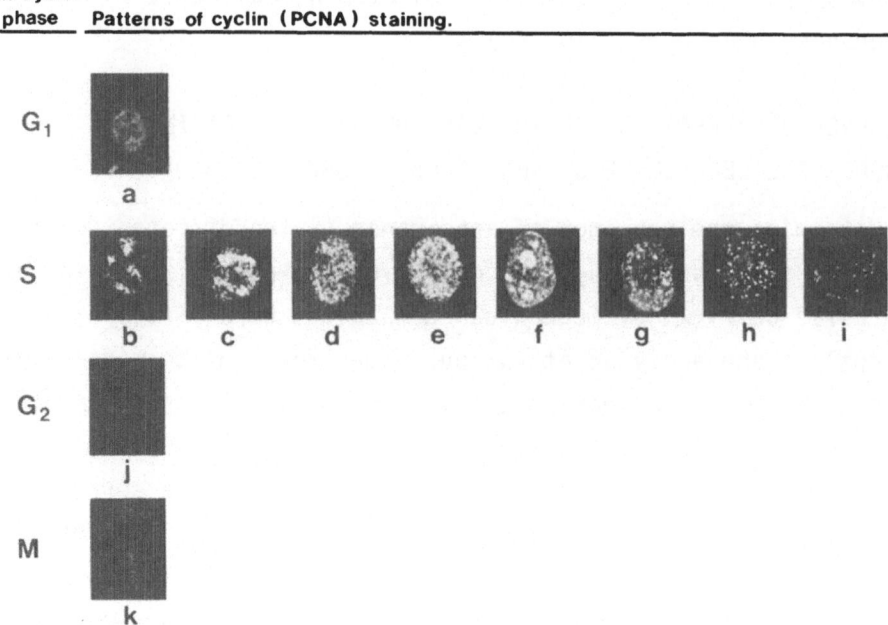

Fig. 4. *Putative sequence of cyclin staining patterns during the cell cycle of AMA cells.* The transition between the different staining patterns does not take place simultaneously in all cells. In most cases, the assignments were aided by the observation of slightly asynchronous sister cells as well as of multinucleated cells showing slightly asynchronous nuclear staining patterns. From Celis and Celis (30).

ing has been demonstrated by comparing phase contrast and immuno-fluorescence micrographs of the same field of cells (see also Figs 2a and b) as well as by double immunofluorescence using a monoclonal antibody that stain these structures (see also Figs 2c and d). A similar, although stronger staining pattern is observed as the cells progress through S-phase (Figs 4c and d). At a later stage, before maximum DNA synthesis (13 to 15 hr after plat-

ing, S-phase, see also Fig. 3), cyclin redistributes to reveal a
punctuated pattern with foci of staining throughout the nucleus
(Fig. 4e). This pattern precedes a major redistribution of
cyclin which is then detected in globular structures that correspond to the nucleolus (Fig. 4f). In some cases, it is also possible to detect cyclin staining in distinct foci located close to
the nuclear membrane (not shown, but see Fig. 5g). At this stage,
the number of cells showing [^{3}H]-thymidine incorporation and
cyclin staining is at or near a maximum (17 to 18 hr in Fig. 3).
Thereafter, there are further changes in the distribution of
cyclin with the pattern becoming punctuated (Fig. 4g) and of decreasing intensity (Figs 4h and i). Cells in G2 (Fig. 4j) and
mitosis (Fig 4k; there is no chromosome staining) show only very
weak staining.

Independent support for the putative sequence of patterns presented in Fig. 4 has been obtained by the analysis of asynchronous
AMA cells labelled with [^{3}H]-thymidine (Fig. 5a) prior to immunofluorescence with PCNA antibodies (Fig. 5b) (30). Some corresponding patterns to those shown in Fig. 4 are indicated with the same
letters in Figs 5a and b. In gereral, we find a good correlation
between the number of grains present per cell and the pattern of
cyclin staining.

Preliminary immunofluorescence studies of early S-phase cells
from human cell lines such as SVK14 (SV40 transformed keratinocytes; 35), A431 (carcinoma of the vulva) and HeLa (cervical
adenocarcinoma) have revealed that the nucleoplasmic staining
pattern exhibiting nucleolar exclusion (see also Figs 4b and c)
is also the first to be detected in these cells (not shown) (30).
Asynchronous cultures of these cell lines (SVK14, Fig. 6a; A431,
Fig. 6b; HeLa, Fig. 6c) exhibit all the nuclear staining patterns
observed in AMA cells. Moreover, these patterns are also detected

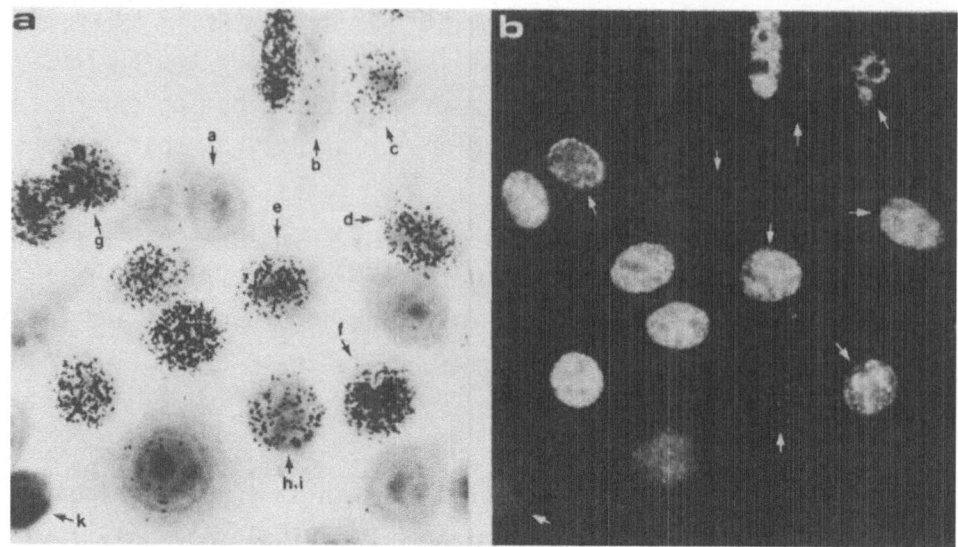

Fig. 5. *Autoradiography (a, [³H]-thymidine incorporation) and epifluorescence (b, PCNA antibodies) of the same field of asynchronous AMA cells.* Cells grown in coverslips were labelled with [³H]-thymidine (30 min. 2 µCi/ml), fixed with methanol and reacted with PCNA antibodies. Immunofluorescence pictures were taken prior to autoradiography. From Celis and Celis (30).

in normal human amnion cells (AF type, Fig. 6d), suggesting that the distribution of cyclin in transformed cells is not a consequence of transformation (30).

SOME PATTERNS OF CYCLIN DISTRIBUTION RESEMBLE THOSE OF DNA REPLICATION

Figs 7a to d show immunofluorescence (PCNA antibodies; Figs 7a and c) and autoradiographic ([³H]-thymidine incorporation; Figs 7b and d) micrographs of homophasic homokaryons produced by PEG induced fusion of mitotic AMA cells which were labelled with [³H]-thymidine and reacted with PCNA antibodies early (Figs 7a and b) and late (Figs 7c and d) during S-phase. Clearly, some of the topographical patterns of DNA synthesis (nucleolar exclusion; Fig. 7b; nucleolar labelling; Fig. 7d) are strikingly similar to those

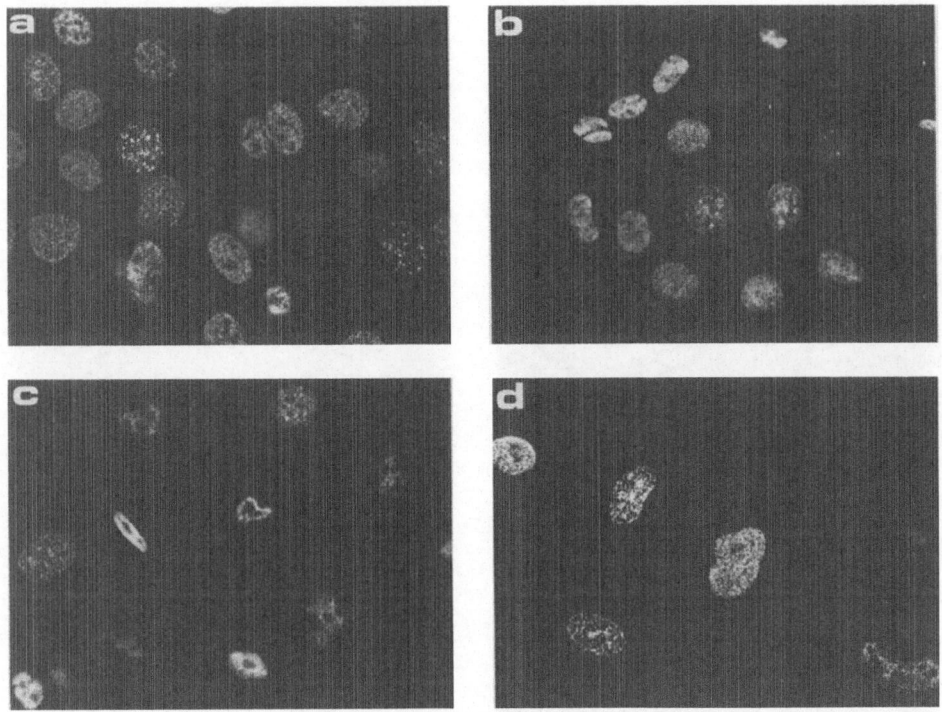

Fig. 6. *Nuclear localization of cyclin in asynchronously growing human cultured cells.* Cells grown in coverslips were treated with methanol prior to immunofluorescence. (a) SVK14, (b) A431, (c) HeLa and (d) human amnion cells, AF type. From Celis and Celis (30).

observed with PCNA antibodies (Figs 7a and b, respectively).

CYCLIN DISTRIBUTION IS DETERMINED BY THE STATUS OF DNA REPLICATION

Evidence suggesting that the distribution of cyclin is determined by the status of DNA replication has been obtained by analyzing the nuclear staining of this protein in synchronized AMA cells treated with inhibitors such as thymidine and hydroxyurea which block cells at the G_1/S border of the cell cycle. Figs 8b (thymidine treated cells; 2 mM) and 8c (hydroxyurea treated cells,

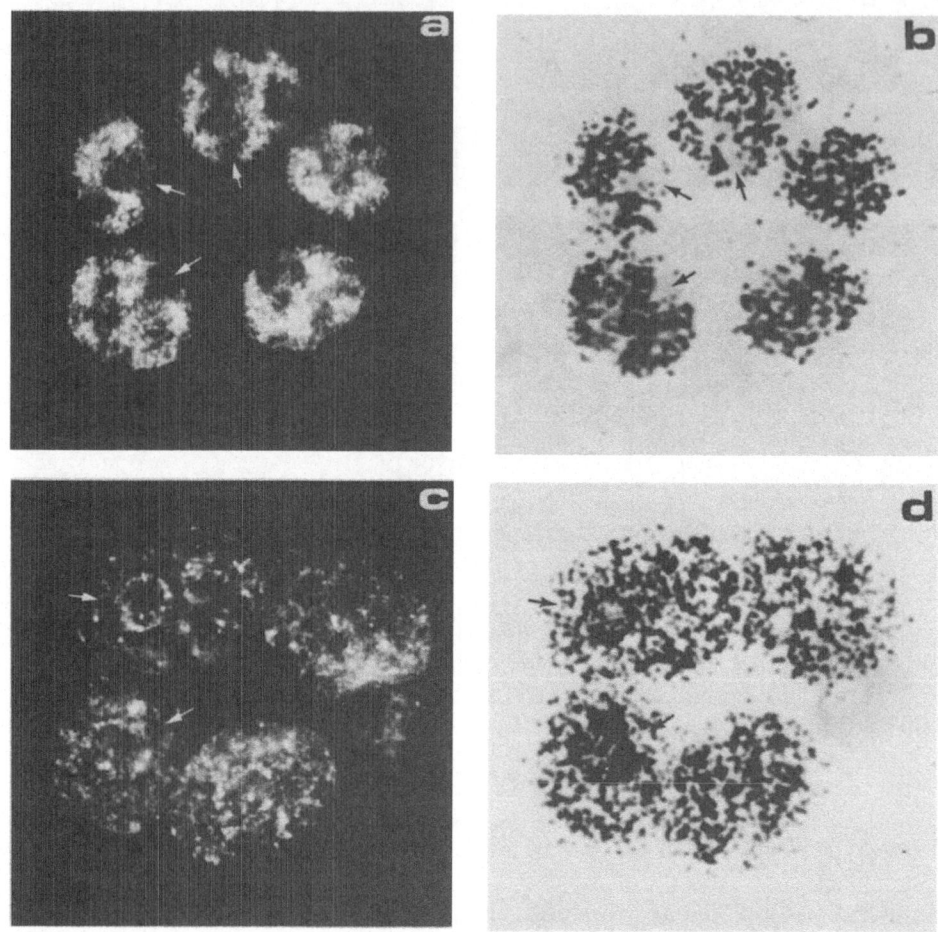

Fig. 7. *Immunofluorescence (a,c; PCNA antibodies) and autoradiography (b,d; [^{3}H]-thymidine incorporation) of the same field of
AMA homophasic homokaryons produced by PEG induced fusion of
mitotic cells.* Homokaryons were labelled with [^{3}H]-thymidine and
processed for immunofluorescence early (a,b) and late (c,d) during S-phase.

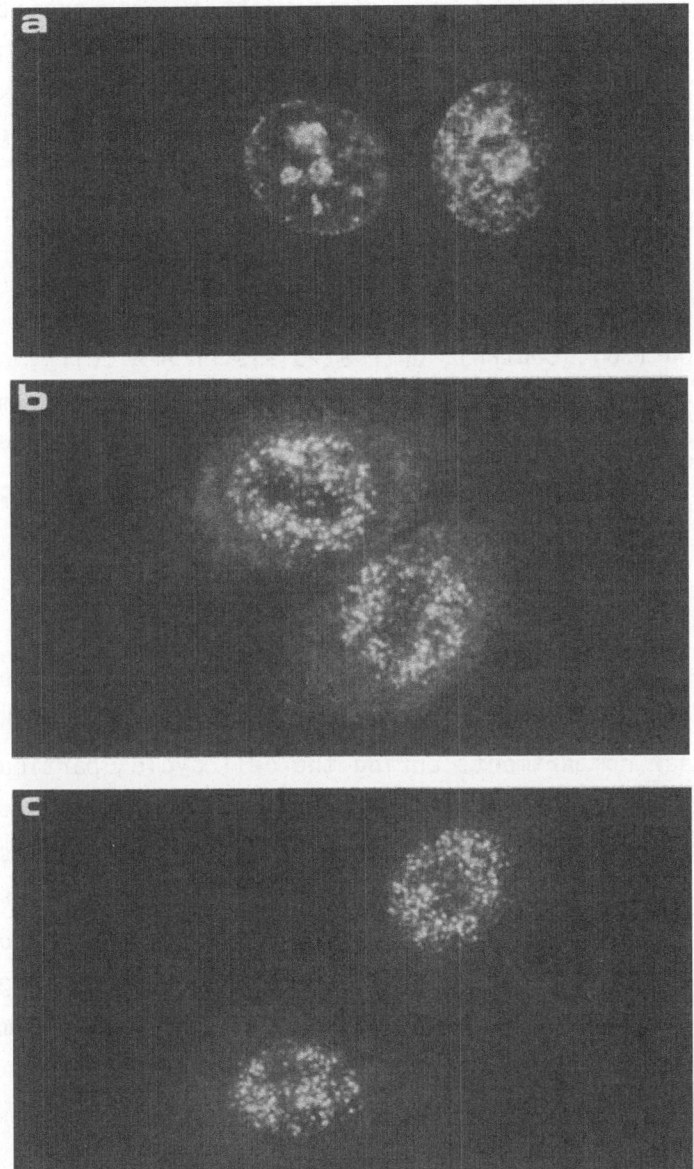

Fig. 8. *Cyclin distribution in synchronous AMA cells treated with thymidine and hydroxyurea.* (a) synchronous sister cells fixed with methanol and reacted with PCNA antibodies 18 hr after plating mitotic cells. (b) as (a) but with thymidine (2 mM) added 1 hr after plating. (c) as (a) but with hydroxyurea (10 mM) added 1 hr after plating.

10 mM) show immunofluorescence micrographs of representative sister AMA cells reacted with PCNÁ antibodies 18 hr after plating
mitotic AMA cells in the presence of the inhibitors. In both cases,
the cells react positively with PCNA antibodies (the staining is,
however, significantly weaker than that observed in control cells;
see Fig. 8a) to reveal a dotted nuclear staining pattern that is
similar to that of early S-phase cells (see also Fig. 4b). The
staining foci must correspond to prereplicative sites of cyclin
localization (36). Control, untreated sister AMA cells, on the
other hand, show mainly nucleolar staining (Fig. 8a). Similar results have been obtained by R. Bravo and H. Mcdonald-Bravo in 3T3
cells treated with hydroxyurea or aphidicolin (personal communication).

CONCLUSIONS

The results presented in this article raise important questions
concerning the mechanism(s) by which cyclin migrates within defined nuclear compartments during the cell cycle, particularly in
S-phase (30). The changes in the distribution of this protein are
dramatic and they have been observed in normal as well as transformed cells of several vertebrate species including aves (chicken
fibroblasts, not shown), bat (lung, CCL 88, Fig. 9a), dog (dog
thymus, not shown), goat (sinovial membrane; growing, Fig. 9b),
mink (lung, CCL 64, Fig. 9c), monkey (TC7, not shown), mouse
(3T3B, Fig. 9d), pisces (blue gill, CCL 91, not shown), rabbit
(cornea, CCL 60, not shown) and sheep (choriodal plexus membrane,
not shown).

There are at least two possibilities that may explain the differential nuclear distribution of cyclin during S-phase: (1) a direct
migration of this protein or (2) migration of a macromolecule(s)
to which cyclin is associated. Conformational changes leading to
masking/demasking of antigenic determinants should also be con-

sidered. Furthermore, the possibility cannot be ruled out that
the observed changes in nuclear staining reflect changes in anti-
genicity only. Further studies will be required in order to distinguish between these possibilities.

At present, little is known concerning the function of cyclin, although from all the available information, it seems likely that
this protein play a key role in some specific aspects of DNA replication. Experiments are now underway to purify cyclin for functional studies as well as to produce monoclonal antibodies that
may be of help to clone the gene. These antibodies should also be
valuable in surgical pathology to assess pre and malignant growth.

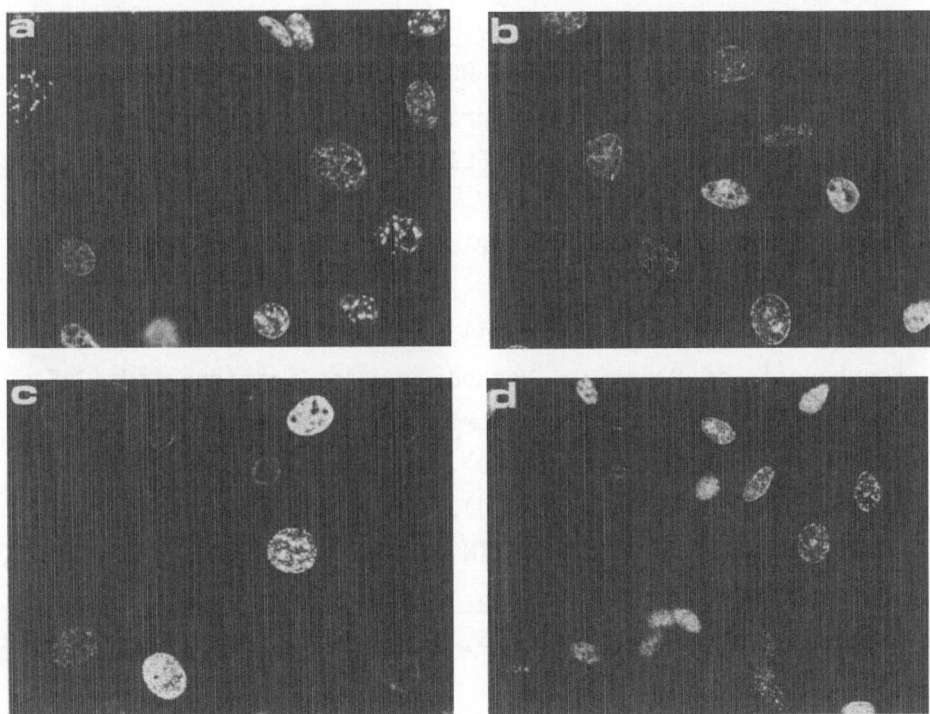

Fig. 9. *Nuclear localization of cyclin in cultured cells from
vertebrate species.* Cells grown in coverslips were treated with
methanol prior to immunofluorescence. (a) bat, (b) goat sinovial
membrane, (c) mink lung and (d) mouse 3T3B. From Celis and Celis
(30).

ACKNOWLEDGEMENTS

We will like to thank Peder Madsen for reading the manuscript and O. Jensen for photography. This work was supported by grants from Euratom, the Danish Medical and Natural Science Research Councils, the Danish Cancer Foundation, Novo, and the Aarhus University Research Fund.

REFERENCES

1) CELIS, J.E., BRAVO, R., MOSE LARSEN, P., FEY, S.J., BELLATIN, J., & CELIS, A. (1984). In: Two-Dimensional Gel Electrophoresis of Proteins: Methods and Applications (eds J.E. Celis & R. Bravo), p. 308, Academic Press, New York.

2) BISHOP, J.M. (1983). Ann. Rev. Biochem., 52, 301.

3) LEVINE, A., TOPP, W., VAN DE WOUDE, G., & WATSON, J.D. eds (1984). Cancer Cells, Cold Spring Harbor Laboratory.

4) BRAVO, R., BELLATIN, J., & CELIS, J.E. (1981). Cell Biol. Int. Rep., 5, 93.

5) BRAVO, R. & CELIS, J.E. (1982). Clin. Chem. (Winston-Salem, NC) 28, 766.

6) BRAVO, R. & CELIS, J.E. (1984). In: Two-Dimensional Gel Electrophoresis of Proteins: Methods and Applications (eds J.E. Celis & R. Bravo), p. 445, Academic Press, New York.

7) BRAVO, R. & CELIS, J.E. (1980). J. Cell Biol., 48, 795.

8) BRAVO, R. & CELIS, J.E. (1980). Exp. Cell Res., 127, 249.

9) BRAVO, R., FEY, S.J., BELLATIN, J., MOSE LARSEN, P., AREVALO, J., & CELIS, J.E. (1981). Exp. Cell Res., 136, 311.

10) BRAVO, R., FEY, S.J., & CELIS, J.E. (1981). Carcinogenesis, 2, 769.

11) CELIS, J.E. & BRAVO, R. (1981). Trends Biochem. Sci., 6, 197.

12) BRAVO, R., FEY, S.J., BELLATIN, J., MOSE LARSEN, P., & CELIS, J.E. (1982). In: Embryonic Development (ed. M. Burger), Part A, p. 235.

13) BRAVO, R. & CELIS, J.E. (1982). Clin. Chem. (Winston-Salem, NC), 28, 949.

14) BELLATIN, J., BRAVO, R., & CELIS, J.E. (1982). Proc. Natl. Acad. Sci. USA, 79, 4367.

15) BRAVO, R., BELLATIN, J., FEY, S.J., MOSE LARSEN, P., & CELIS, J.E. (1983). In: Gene Expression in Normal and Transformed Cells (eds J.E. Celis & R. Bravo), p. 263, Plenum, New York.

16) FORCHHAMMER, J. & MACDONALD-BRAVO, H. (1983). In: Gene Expression in Normal and Transformed Cells (eds J.E. Celis & R. Bravo), p. 291, Plenum, New York.

17) CELIS, J.E. & BRAVO, R. (1984). FEBS Lett., 165, 21.

18) BRAVO, R. (1984). FEBS Lett., 169, 185.

19) CELIS, J.E., BRAVO, R., MOSE LARSEN, P., & FEY, S.J. (1984). Leuk. Res., 8, 143.

20) CELIS, J.E., FEY, S.J., MOSE LARSEN, P., & CELIS, A. (1984). Proc. Natl. Acad. Sci. USA, 81, 3128.

21) MATHEWS, M.B., BERNSTEIN, R.M., FRANZA, R., & GARRELS, J.I. (1984). Nature, 309, 374.

22) CELIS, J.E., FEY, S.J., MOSE LARSEN, P., & CELIS, A. (1984). Cancer Cells, 1, 123.

23) FRANZA, B.R. & GARRELS, J.I. (1984). Cancer Cells, 1, 137.

24) BRAVO, R. (1984). Cancer Cells, 1, 147.

25) BRAVO, R. (1984). Proc. Natl. Acad. Sci. USA, 81, 4848.

26) MIYACHI, K., FRITZLER, M.J., & TAN, E.M. (1978). J. Immunol., 121, 2228.

27) TAKASAKI, Y., DENG, J.S., & TAN, E.M. (1981). J. Exp. Med., 154, 1899.

28) TAN, E.M. (1982). Adv. Immunol., 33, 167.

29) TAKASAKI, Y., FISCHWILD, D., & TAN, E.M. (1984). J. Exp. Med., 159, 981.

30) CELIS, J.E. & CELIS, A. (1985). Proc. Natl. Acad. Sci. USA, in press.

31) CELIS, J.E. & BRAVO, R. (1984). In: Electrophoresis '84 (ed.
    V. Neuhoff), p. 205, Verlag Chemie.

32) WILLIAMS, C.A. & OCKEY, C.H. (1970). Exp. Cell Res., 63, 365.

33) YANISHEVSKY, R.M. & PRESCOTT, D.M. (1978). Proc. Natl. Acad.
    Sci USA, 75, 3307.

34) CELIS, J.E. & CELIS, A., in preparation.

35) TAYLOR-PAPADIMITRIOU, J., PURKIS, P., LANE, E.B., MCKAY, I.A.,
    & CHANG, S.E. (1982). Cell Differ., 11, 169.

36) QUINLAN, M.P., BO CHEN, L., & KNIPE, D.M. (1984). Cell, 36,
    857.

REGULATION OF GENE EXPRESSION IN DEVELOPMENTAL AND ONCOGENIC PROCESSES:

THE ALBUMIN ALPHA-FETOPROTEIN LOCUS IN MAMMALS

*José M. Sala-Trepat, Anne Poliard, Isabelle Tratner,
Maryse Poiret, Mariela Gomez-Garcia, Andras Gal,
Jean-Louis Nahon & Monique Frain*

Laboratoire d'Enzymologie, C.N.R.S.
91190 Gif-sur-Yvette, France

INTRODUCTION

The molecular mechanisms underlying malignant cell transformation
are hardly understood. An attractive hypothesis is that the acquisition of the malignant phenotype might be brought about by the
activation of genes (e.g. oncogenes) whose expression is normally
restricted to actively proliferating embryonic cells. Insight into
the mechanisms controlling gene expression during developmental
and oncogenic processes appears then essential to our understanding of neoplastic transformation.

The control of albumin and alpha-fetoprotein (AFP) synthesis
in the mammalian liver provides a valuable model system to
investigate this problem. Indeed, these two serum proteins show a
reciprocal relationship in their plasma levels during normal
oncogenic development and in some oncogenic events (1, 2). Thus,
AFP is the major plasma protein of the mammalian fetus, where it
is synthesized by the developing liver and the yolk sac (3, 4).
The concentration of this protein is drastically decreased in the
serum of the normal adult (1, 4). On the contrary, albumin is the

dominant plasma protein during adult life and its concentration
increases from low levels early in fetal development to high,
approximately constant, values in postnatal life (1, 5). The
parenchymal cells of the liver are the main site of synthesis of
this protein during both embryonic and adult life (6). In addition
albumin is produced by the yolk sac of certain species (human,
mouse and chick) but not by the rat yolk sac (6, 7).

AFP synthesis by the adult liver can be resumed under certain
physiopathological conditions leading to restitutive cell
proliferation such as regeneration of the liver following partial
hepatectomy and chemically induced liver necrosis (1, 4, 8).
Increased levels of AFP in the serum can also be observed during
the preneoplastic stages of liver carcinogenesis (2, 9, 10). In
all these cases alterations of serum AFP are, however, relatively
small. Highly elevated plasma levels of AFP in the adult are
generally associated with the appearance of tumors arising from
liver cells and yolk sac elements, in analogy with the embryonal
sites of AFP synthesis (1, 2, 4). In most of the processes which
lead to AFP reexpression, albumin synthesis is diminished as
compared to that of normal liver. For instance, most transplantable rat hepatomas which produce high levels of AFP show a much
reduced rate of albumin synthesis (11-13).

These changes in cell-specific protein synthesis provide an
attractive model for investigating the regulatory mechanisms
involved in the activation and inactivation of specific genes
during embryonic development and neoplastic transformation. In the
first part of this article, we briefly review present knowledge on
the structure of the albumin and AFP genes, and summarize
experiments indicating that these two genes are regulated at the
transcriptional level. We then describe data from our laboratory
showing that modifications at the genomic DNA level (such as

amplification-deletion, rearrangements, changes in location of
gene sequences within the supercoiled DNA loops or changes in
methylation pattern) do not appear to be involved in the
transcriptional modulation of these genes. Finally, we summarize
our recent results on the chromatin structure of the albumin and
AFP genes in developing rat liver and other rat tissues and cell
lines.

STRUCTURE OF THE ALBUMIN AND ALPHA-FETOPROTEIN GENES

In recent years, rapid progress has been made in the analysis
of the structure of the mammalian albumin and AFP genes. First,
the isolation and purification of the albumin and AFP mRNA
molecules from rat and mouse (14-17) made possible the cloning of
the corresponding cDNAs (18-22). The cloned cDNA probes were used
to identify the corresponding genomic sequences in rat and mouse
DNA libraries (19, 21, 23). Characterization of these cDNA and
genomic clones revealed that both genes are split and organized
similarly into 15 coding fragments (exons) interrupted by 14
intronic sequences (23, 24). The internal 12 exons of either gene
are composed of three similar sets of 4 exons (24, 25). This
provides a genetic basis for the three-domain structure of albumin
and AFP (26). Nucleotide sequence homologies among the four exons
that constitute a single domain also suggest that both genes arose
from a common sequence which underwent successive amplification
and divergence (24, 25, 27). In addition, it has been shown that
in the mouse the albumin and AFP genes are closely linked in
tandem on chromosome 5, with the 3'-terminus of the albumin gene
preceding at 13.5 Kb the 5' site of the AFP gene (28). In the rat
both genes have been shown to be syntenic on chromosome 14 (29).
Recently, cloning of human albumin and AFP cDNAs has also been
accomplished and the structure of the human genes is being
investigated (30-32). Both albumin and AFP genes have been found
to map within bands q11-22 of the long arm of human chromosome 4

(33). The close linkage of the albumin and AFP genes in the mouse, and probably in other mammalian species, suggest the existence of a single functional locus which would be under coordinate control during development.

TRANSCRIPTIONAL CONTROL OF THE ALBUMIN AND ALPHA-FETOPROTEIN GENES DURING NORMAL DEVELOPMENT AND NEOPLASIA

Studies on the regulation of the albumin and AFP genes during development were first carried out by solution hybridization analysis using uncloned single-stranded cDNA probes. Determination of the steady-state levels of albumin and AFP mRNAs in polysomal RNA preparations from rat liver at different stages of development and from different rat hepatomas indicated the existence of a close correlation between the concentration of mRNA sequences and the specific albumin and AFP protein synthetic activities (10, 12-14, 34-36). The developmental changes in albumin and AFP mRNA sequences observed in these studies were confirmed by Dot hybridization analysis using total RNA preparations and cloned albumin and AFP cDNA probes (Fig. 1). These results clearly established that the expression of the albumin and AFP genes in developing rat liver and in the different hepatomas studied is mainly regulated by modulating the steady-state concentrations of the corresponding functional mRNAs rather than by translational control mechanisms.

A detailed analysis of the subcellular distribution of albumin and AFP mRNA sequences in developing rat liver and in the Morris hepatoma 7777 has indicated that, in all cases, most of the albumin and AFP mRNA sequences are found associated with the polysomes as mature mRNA molecules (37) ; no evidence was found for storage of inactive mRNA sequences in the nuclear or cytoplasmic extra-polysomal compartments in any stage of liver development or in the hepatoma tissue (37). These results argue

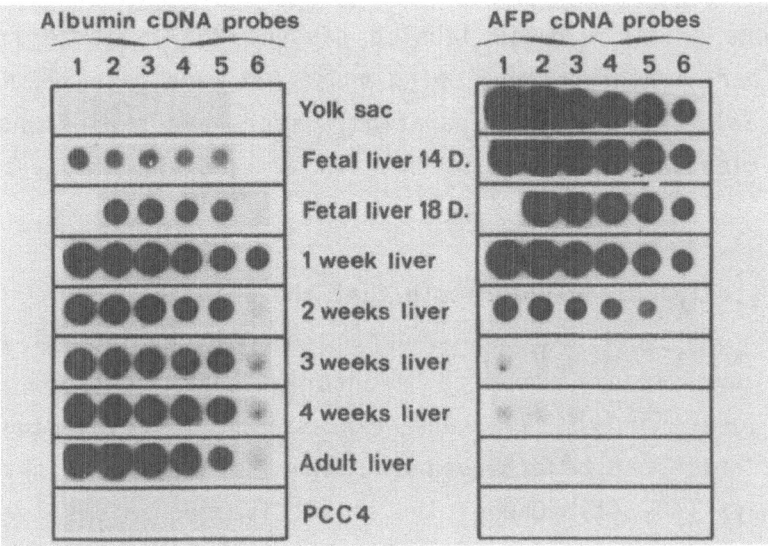

Figure 1. *Developmental changes in albumin and AFP mRNA sequences in total cell RNA from rat liver and yolk sac.* Dot blot hybridizations with the ³²P-labeled albumin and AFP cloned rat cDNA probes are shown. 1 to 6, decreasing amounts of RNA (4 to 0.1 μg). RNA samples from the mouse embryonal carcinoma cell line PCC4 were included as control for background hybridization.

for an extremely efficient processing of the primary transcripts of the albumin and AFP genes to mature mRNAs which would be rapidly released to the polysomes to be utilized for protein synthesis. These data thus provided no indication that post-transcriptional mechanisms might play an important role in the developmental regulation of the albumin and AFP genes ; they rather suggest that the expression of these genes during normal development and neoplasia is controlled at the transcriptional level.

Direct evidence showing that the decrease in AFP mRNA molecules after birth (see Fig. 1) is a result of a much reduced transcription of the AFP gene has been obtained by cell-free nuclear transcription experiments. The amounts of specific albumin and AFP gene transcripts in labeled nascent RNA isolated from nuclei preparations from developing mouse (38) and rat (39, Nahon, Danan and Sala-Trepat, in preparation) liver have been found to correlate closely with those of the albumin and AFP steady-state mRNAs.

All these studies clearly indicate that the changes in albumin and AFP gene expression during normal development and neoplasia are mainly due to transcriptional control. Transcriptional control of gene expression in developing eukaryotic cells can be achieved by mechanisms acting in several possible ways (40, 41). One of the attractive mechanisms is that some differential change on a specific part of DNA itself, like a DNA rearrangement, selective amplification or deletion of certain gene sequences, or base modifications could be the cause of selective gene activation or "repression" during development.

INVARIANT ORGANIZATION OF THE ALBUMIN AND ALPHA-FETOPROTEIN GENES IN FETAL AND ADULT RAT TISSUES AND RAT HEPATOMAS

There are now several examples in which rearrangements of DNA are involved in the regulation of gene expression in higher eukaryotes. In particular, rearrangements are required for the activation of immunoglobulin genes during normal B-lymphocyte differentiation (42) and the genesis of B-cell-derived tumors in mice and men (43). We have investigated whether DNA rearrangements accompany the changes in transcriptional activity of the albumin and AFP genes that occur during rat liver development and neoplasia. This was done by using the Southern blotting technique to compare the organization of the albumin and AFP genes in fetal, newborn and adult rat tissues (liver, kidney and spleen) and in two rat hepatomas (7777, 8994) which show drastic differences in the level of expression of these two genes (13, 14, 37, Nahon *et al.*, submitted).

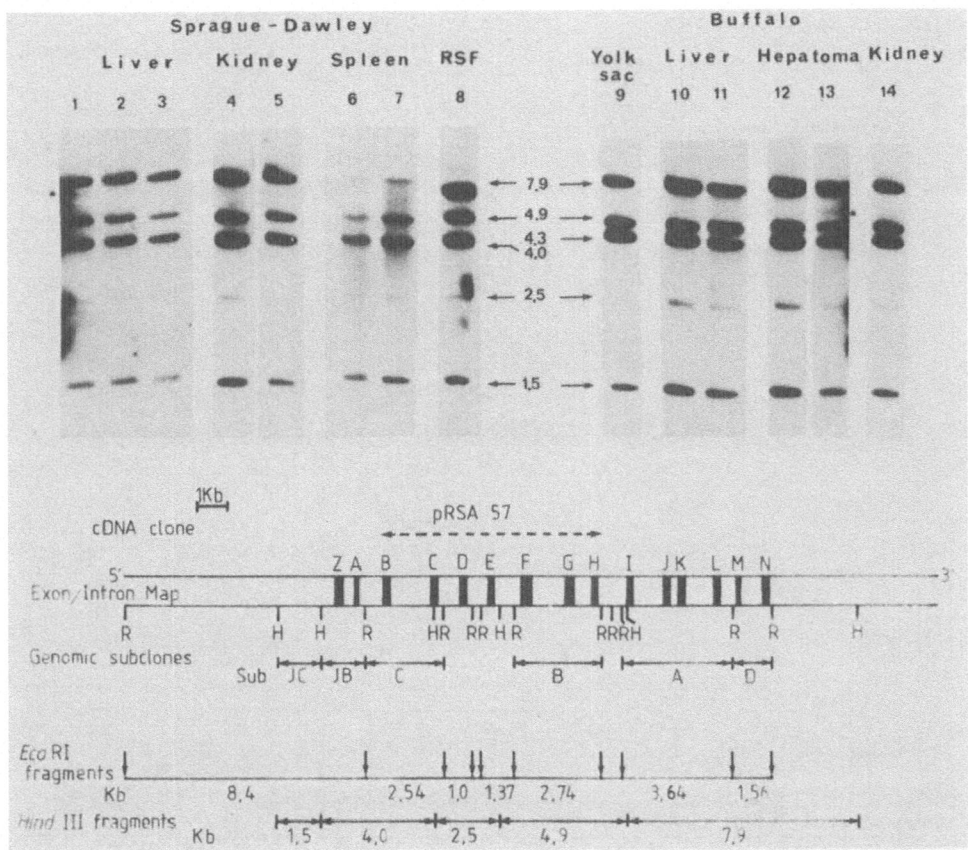

Fig. 2. *Identification of Hind III restriction fragments containing intragenic and extragenic sequences of the albumin gene in
chromosomal DNA from different rat tissues and hepatomas.* Autoradiographs of rat DNA from different tissues of the Sprague-
Dawley and Buffalo strains and from two Morris hepatomas after
Hind III digestion, gel fractionation and Southern blot hybridization with a mixture of the ^{32}P-labeled albumin and genomic probes (pRSA 57, sub JC, sub JB, sub C, sub B, sub A and sub D). The
physical map of the Sprague-Dawley rat albumin gene is shown in
the lower part of the Figure. Black vertical boxes denote exons,
and white boxes introns. R, EcoR I sites. H, Hind III sites.

Southern blots of DNAs digested with one or another of the
restriction endonucleases Eco RI, Hind III or Msp I, were hybridized to albumin and AFP cDNA and genomic probes. As can be seen

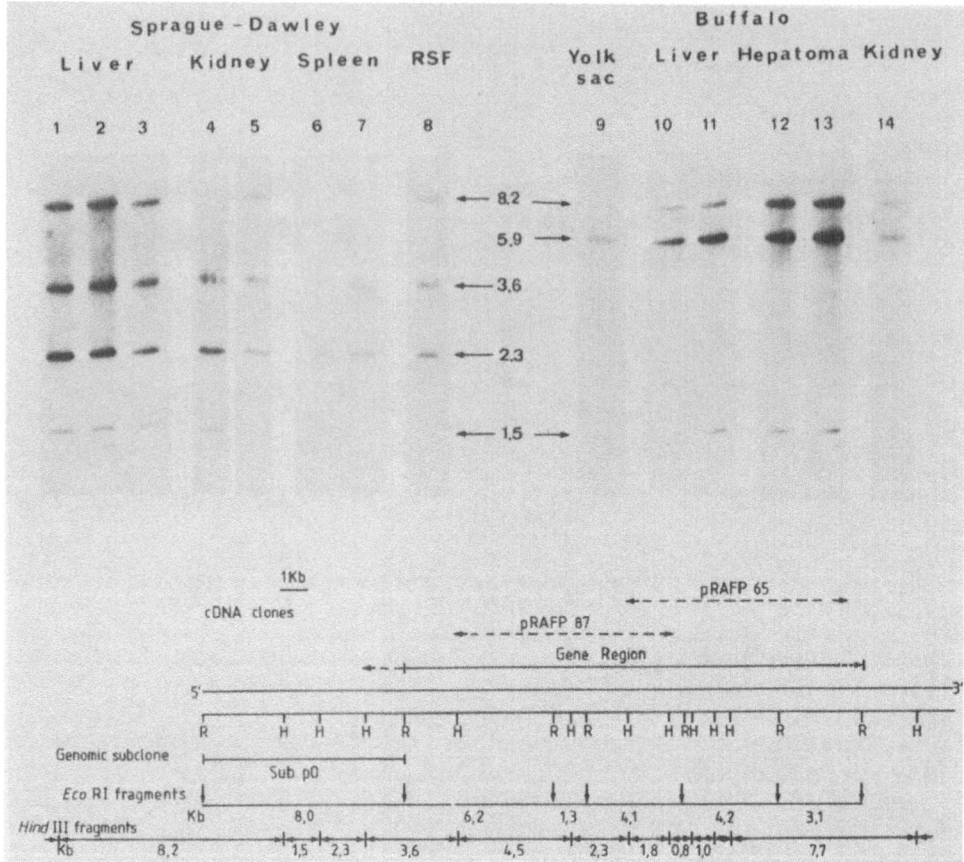

Fig. 3. *Comparison of the Hind III restriction fragments corresponding to the 5'end and flanking region of the AFP gene in chromosomal DNA from different rat tissues and hepatomas.* Experimental conditions were as indicated in the legend of Fig. 2, except that the DNA fragments fixed on the nitrocellulose filters were hybridize to the AFP genomic subclone pO. The restriction map of the Sprague-Dawley rat AFP gene is shown schematically below the autoradiographs. R, Eco RI sites. H, Hind III sites.

from Fig. 2 and Fig. 3 for the Hind III digestions, there are no apparent differences in the hybridization patterns of the chromosomal DNAs from the variety of tissues we have analysed. In contrast, differences can be observed in the restriction patterns of both albumin (Fig. 2) and AFP genes (Fig. 3 and not shown) in the DNAs isolated from Sprague-Dawley and Buffalo rats. This is in line with recent results from our laboratory that have shown the existence of extensive allelic polymorphism between these two inbred strains of rats (23, 44, 45). The same kind of observations have been made following Eco RI or Msp I digestions of the DNAs and hybridization to the different albumin and AFP cDNA and genomic probes, which permit to compare not only the gene regions but also the flanking sequences extending 7 to 10 Kb usptream from the 5' ends of these genes and 3 Kb downstream from the 3'-ends (see gene maps on Figs 2 and 3 ; Gal *et al.*, submitted).

All these results indicate that the albumin and AFP genes are not grossly rearranged during development and neoplastic trans-formation ; they appear to remain invariant throughout the regu-latory processes involved in their tissue- and time-specific transcription. Our studies have further shown (Gal *et al.*, sub-mitted) that these genes are present only once in the haploid genome of all the tissues and hepatomas analysed, unless the entire 27 Kb and 35 Kb genomic units of the albumin and AFP genes shown in Fig. 2 and Fig. 3, respectively, are repeated a few times. There is thus no evidence for amplification or deletion of these genes during normal development and neoplastic transformation.

IS CLOSE ASSOCIATION OF THE ALBUMIN AND ALPHA-FETOPROTEIN GENES WITH THE NUCLEAR CAGE A NECESSARY CONDITION FOR THEIR ACTIVE TRANSCRIPTION?

It is now widely accepted that at higher structural level, nuclear DNA is organized into loops (46,47) by attachment to the so-called nuclear "matrix" (48) or nuclear "cage" (46), a structure which has been implicated in transcription and processing of RNA (49,50). The specific association of actively

transcribed genes with the nuclear cage has been reported for some
viral and eukaryotic genes (49, 51-52). This association has been
hypothesized to be implicated in the transcription of tissue-
specific genes. To examine whether the transcriptional control of
albumin and AFP genes is mediated through events involving close
attachment to the nuclear cage, we have analysed the relative
location of these genes with respect to the nuclear cage in
hepatic and fibroblastic cell lines. We first isolated "nucleoids"
(DNA replete nuclear cages ; 53) from the 7777-C8 rat hepatoma
cell line and the JF-1 fibroblasts line (54, 55) by lysing the
cells in a detergent and 2M NaCl, and sedimenting the loop
structures through a step gradient as described by Cook and
Brazell (53). These "nucleoid" structures retained normal nuclear
morphology as judged by phase contrast and electron microscopy.
The nucleoids were partially digested with the restriction
endonuclease EcoRI such that DNA is gradually cleaved from the
cage. The cages with the remaining attached DNA were sedimented
free from unattached DNA which was recovered in the supernatant
(53). The attached (5-15 % of the total) and unattached DNA was
purified, redigested to completion with EcoRI and analysed by the
presence of specific albumin and AFP gene sequences by the method
of Southern. The results are shown in Fig. 4.

The pattern of hybridization and the intensity of the bands
of the matrix DNA samples containing 15 % of total nuclear DNA is
not significantly different from that of DNA isolated from total
undigested nucleoids, in none of the two cell lines. Further, no
significant differences are observed in the hybridization patterns
of the cage attached and unattached DNA sequences in digestions in
which only 5 % of the total DNA remained bound to the nuclear cage
(see Fig. 4). These results indicate that DNA lying close to the
nuclear cage is not enriched (or depleted) in specific AFP (or
albumin) gene sequences in the hepatoma cells. In both cell lines

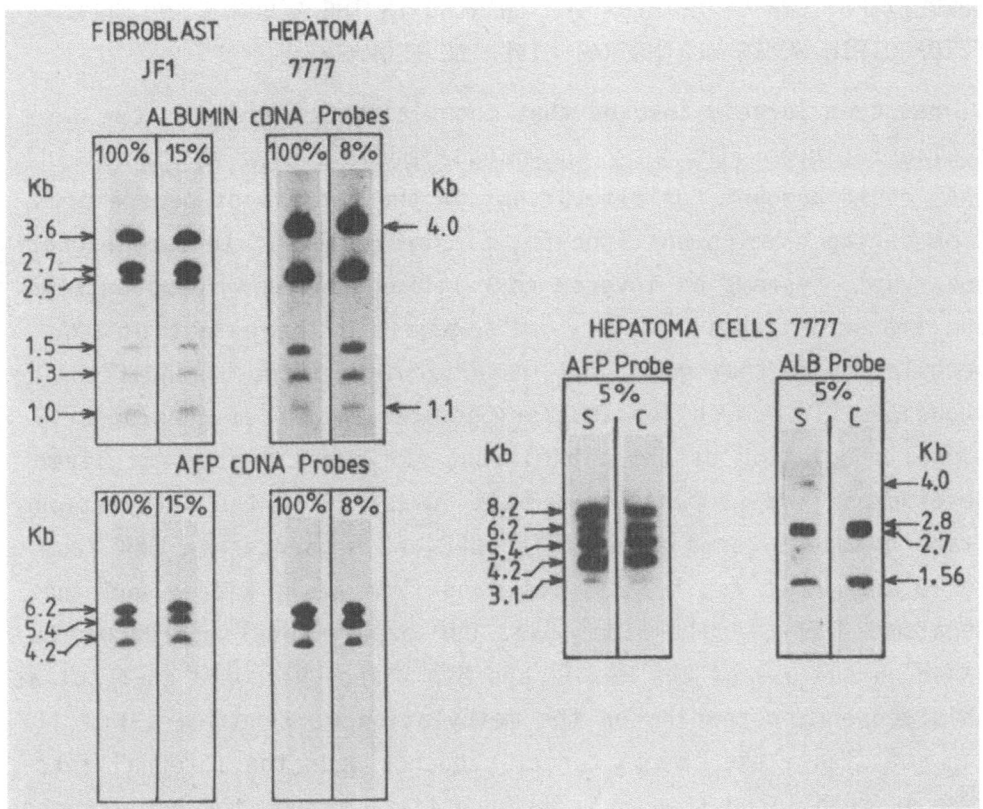

Fig. 4. *Albumin and AFP gene fragments in total DNA (100 %) and nuclear cage-associated DNA (15 %, 8 % or 5 %) following Eco RI cleavage.* Total nucleoids were prepared and partially cleaved with Eco RI as indicated in the text. The DNA from pelleted nuclear cages (C) and the unattached DNA in the supernatant fraction (S) were isolated and digested to completion with Eco RI. Equal amounts of DNA (8 µg) were fractionated on 0.8 % agarose gels, transferred to nitrocellulose filters and hybridized to cloned albumin or AFP cDNA probes.

the AFP and albumin gene sequences appear to be randomly distributed with respect to the nuclear cage. As the 7777-C8 cell line actively transcribes the AFP but not the albumin gene (53) while in the fibroblast line both genes are silent, our studies show that preferential association of the albumin and AFP genes with the nuclear cage is not a necessary requirement for their active transcription.

COMPLEX PATTERN OF METHYLATION CHANGES IN THE ALBUMIN AND ALPHA-
FETOPROTEIN GENES DURING RAT LIVER DEVELOPMENT

It has been largely invoked that methylation of specific cytosine residues (the only known base modification of mammalian
DNA) could account for alterations in the pattern of gene expression during development (see 56, 57 for reviews). In many higher
eukaryotic systems an inverse correlation has been found between
the transcriptional activity of a gene and the extent of DNA
methylation of that gene, and in particular of the 5'-end flanking
sequences. In search for another possible mechanism of transcriptional modulation of the albumin and AFP genes during rat liver
development and neoplasia, we have investigated the methylation
state of these genes at specific C-C-G-G sequences in DNA from
fetal and adult rat hepatocytes, and from adult kidney and the
hepatoma 7777-C8 cell line (54). Our experimental approach is
based on the use of the Hpa II and Msp I isoschizomers that cut at
this sequence depending on the methylation of cytosine ; Hpa II
can not cleave the common target sequence when the internal cytosine is methylated (see 57). We have first mapped the Msp I sites
in the rat albumin and AFP genes by restriction analysis of appropriate genomic subclones. The ten Msp I sites identified in the
rat albumin gene are distributed across the entire transcription
unit (see Fig. 5), allowing us to sample methylation of the gene

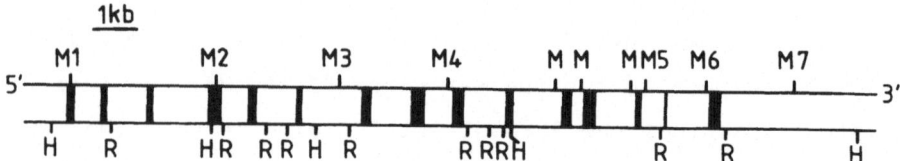

Fig. 5. *Msp I sites identified in the rat albumin gene.* Black
boxes represent coding sequences and white boxes intervening
sequences. M1 to M7 are the Msp I sites whose methylation has been
measured. The three sites denoted by M have not been analysed.

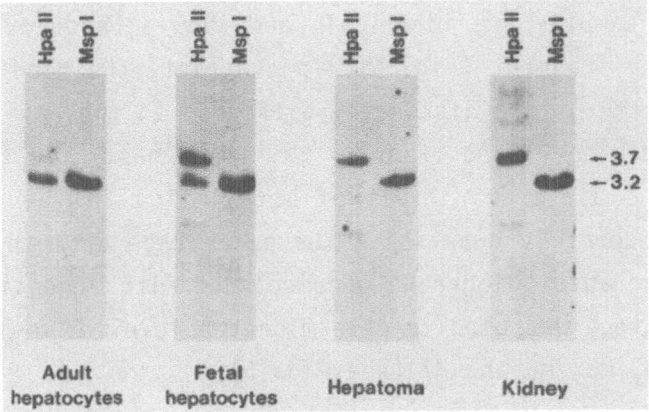

Fig. 6. *Analysis of the level of methylation at the M1 site of the albumin gene in fetal, adult and neoplastic rat cells.* The DNA from the different cells and tissues as indicated, was cleaved with the restriction endonucleases Hind III and either Hpa II or Msp I in succession. The ensuing fragments were fractionated on a 0.8 % agarose gel and hybridized to the rat albumin gene genomic subclone sub C (see left part of the Figure). The autoradiographs of the Hind III-Hpa II and Hind III-Msp I double digestions are shown on the right part of the figure.

from the 5' to the 3' end. Similarly the eight Msp I sites detected in the rat AFP gene span all the different regions of the gene (5', middle and 3' ; not shown).

The extent of methylation of six of the ten Msp I sites of the albumin gene was measured by obtaining large restriction fragments containing the potentially methylated Hpa II sites, and digesting then with Hpa II followed by Southern blot hybridization to appropriate subgenomic molecular probes. For instance, the methylation of the M1 site was studied by hybridizing the fragments resulting from the Hind III-Msp I and Hind III-Hpa II double digestions of the DNAs to the genomic subclone C (see Fig. 6). Cleavage of the 3.7 Kb Hind III fragment located at the 5'-end of the albumin gene at site M1 yields a 3.2 Kb double digestion fragment. As expected the Hind III-Msp I patterns observed with all the DNA samples are identical, and show a band corresponding to the 3.2 Kb fragment (Fig. 6). In contrast, the Hind III-Hpa II digestion patterns are distinctly different in the hepatocytes and hepatoma and kidney DNAs (Fig. 6) ; this demonstrates that the M1 site presents different levels of methylation in the different cells. To accurately quantitate the percentage of methylation at this site the autoradiographs were scanned with a densitometer. A similar strategy was used to determine the level of methylation of sites M3, M4, M5, M6 and M7 (see Fig. 5).

The results of all these studies are summarized in Table 1 which also presents the level of albumin gene expression in the different cells and tissues. All six sites are totally methylated in adult kidney and almost fully unmethylated in adult hepatocytes. In the hepatoma cells these sites were completely methylated except for site M3 which present a 50 % methylation level. These results show a strong correlation between gene activity and undermethylation as has been found for other genes coding for tissue-specific proteins (see 56, 57) and genes turned

Table 1. *Methylation of the albumin gene in different rat cells and tissues.*

Cells and Tissues	Percent Methylation of Site						Albumin gene expression (Number of albumin mRNA molecules per cell)
	M1	M3	M4	M5	M6	M7	
Fetal Hepatocytes	65	100	100	80	90	90	16,000
Adult Hepatocytes	10	10	10	10	10	10	22,000
Hepatoma 7777-C8 cell line	100	50	100	100	100	90	< 30
Adult Kidney	100	100	100	100	100	100	5 - 10

off after malignant transformation (58). However, in 19-day fetal
hepatocytes which actively transcribe the albumin gene all these
sites are highly methylated with site M1 presenting a methylation
level of 65 %. This methylation pattern does not support a
straight correlation between hypomethylation and gene activity.
Since it has been shown that only the undermethylation of the
5'-end M1 site is necessary for albumin gene expression in different rat hepatoma lines (59), it was of interest to determine
whether the partial methylation found at this site in 19-day fetal
hepatocytes could be attributed to functional heterogeneity in the
cell population. *In situ* hybridization studies with the isolated
fetal hepatocytes have shown that not less than 80 % of the isolated cells were labeled by specific albumin cDNA probes (Poliard
et al., in preparation). It is possible that only one of the
chromosomes within the 19-day fetal hepatocytes is methylated at
this site, thus reflecting the developmental stage at which the
undermethylation of this site is being established. However, since
the 16-day fetal hepatocytes already show active transcription of
the albumin gene (see Fig. 1) our results argue against the
possibility that the methylation state of the M1 site determines
whether or not the albumin gene will be expressed during
development.

The methylation state of the AFP gene in the different rat
cells and tissues was analysed following HpaII or MspI digestions
of DNA samples, Southern blotting and hybridization to the cloned
rat AFP cDNA probes pRAFP 65 and pRAFP 87 (see Fig. 3). The
results presented in Fig. 6 show that the AFP gene is highly
methylated in adult kidney. The patterns obtained for fetal and
adult hepatocytes are indicative of extensive, though partial,
methylation of C-C-G-G sequences over the entire AFP gene region.
The only significative difference between the two patterns is the

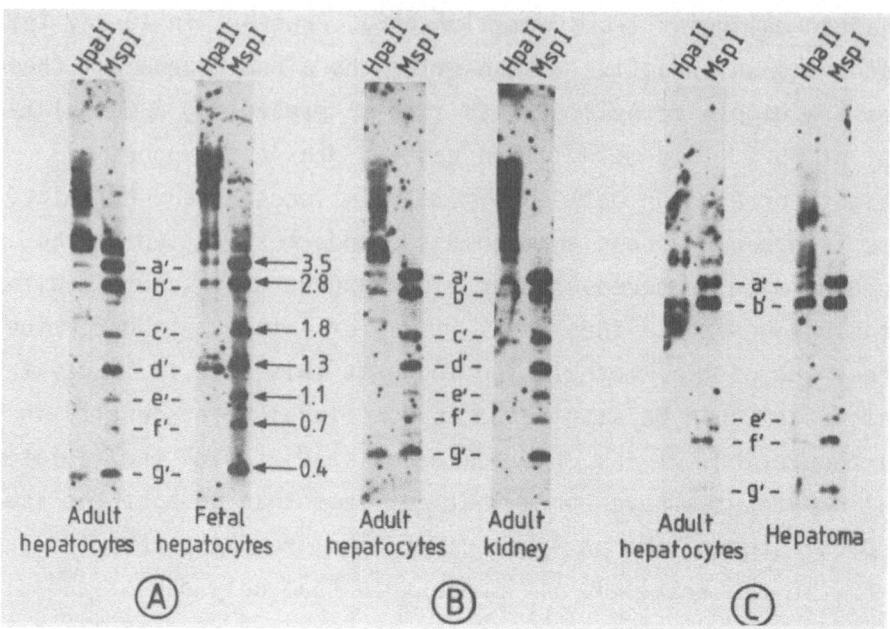

Fig. 7. *Methylation pattern of C-C-G-G sequences in the AFP gene in different rat cells and tissues.* DNA from the different cell populations was digested either with Hpa II or Msp I and the fragments were electrophoresed on 0.8 % agarose gels and analysed by Southern blot hybridization to the AFP cDNA probes pRAFP 65 and pRAFP 87 (20).

presence of the g' fragment in the DNA sample of adult hepatocytes after Hpa II digestion ; this band is not present in the Hpa II digestions of fetal hepatocytes. This observation indicates that the corresponding MspI sites located in the central region of the AFP gene (not shown) are highly methylated in fetal hepatocytes while they are unmethylated in the adult state.

Taken together, these data show that specific changes in the level of methylation of the albumin and AFP genes take place during rat liver development. However, the relationship between these changes in methylation pattern and gene activity are complex. For instance, the albumin gene is highly methylated in 19-day fetal hepatocytes and hypomethylated in adult hepatocytes, though no important differences in the transcriptional state of this gene are observed at these two stages of development (Table 1). In contrast, the extent of methylation of the AFP gene is somewhat higher in fetal than in adult hepatocytes, while the adult hepatocytes show a much reduced transcription of the AFP gene. Similar results have been obtained by Kunnath and Locker (60). All these findings then indicate that alterations in the methylation pattern of the albumin and AFP intragenic sequences do not seem to play a major role in modulating the transcriptional activity of these genes during rat liver development.

However, these studies do not exclude the possible existence of critical methylation sites in the 5'-end flanking sequences of the albumin and AFP genes that could determine the selective expression of these genes. We have searched for such hypothetical sites by studying the level of methylation of specific C-C-G-G sequences extending 4 Kb upstream from the albumin and AFP genes. We have found no rigidly consistent configuration of methylation of a specific site that invariably correlates with gene expression in these flanking regions (Tratner and Sala-Trepat, unpublished data).

CHROMATIN STRUCTURE OF THE ALBUMIN AND ALPHA-FETOPROTEIN GENES DURING DIFFERENTIAL GENE EXPRESSION

A large body of evidence suggests that changes in the chromatin structure of specific genes may play an important role in the transcriptional regulation of eukaryotic genes. It has become clear since the work of Weintraub and Groudine (61) that active genes have altered chromatin structures which render them preferentially sensitive to digestion by the endonuclease DNase I (reviewed in 41 and 62). In this context, it was of interest to

investigate whether alterations in the chromatin structure of the
albumin and AFP genes are associated with the changes in
expression of these genes during development and neoplasia.

As a first approach, this question was analysed by studying
the overall DNase I sensitivity of the albumin and AFP genes in
newborn and adult liver, and in adult kidney. It has been found
that the chromatin regions containing the albumin and AFP genes
are much more sensitive to the nucleolytic action of DNase I in
adult liver than in adult kidney (55). Both albumin and AFP genes
appear to be very sensitive to DNase I in adult liver. In newborn
liver the level of sensitivity of the albumin and AFP genes is not
significantly different from that found in adult liver (Nahon and
Sala-Trepat, unpublished data). Although the AFP gene is not
significantly transcribed in adult hepatocytes, the high degree of
DNase I sensitivity of the AFP gene in these cells might reflect
the fact that it has been previously actively transcribed or it
might indicate its potential for reexpression in oncogenic
processes (see 62). Whatever the significance of this observation,
these results taken together indicate that : a) alterations in the
chromatin structure of the albumin and AFP genes might be involved
in the early establishment of the tissue-specific potential of
overt gene expression ; b) such alterations reflected in an
altered overall DNase I sensitivity do not appear to be
responsible for the changes in gene activity occurring during the
terminal differentiation of the hepatocyte.

In a number of actively transcribed genes, it was found that
in addition to the preferential DNase I sensitivity of the entire
gene region, there exists small regions of nuclease hypersensitivity usually located 5' to the coding region (63, 64). We have

recently searched for the presence of DNase I-hypersensitive
cleavage sites in the chromatin regions flanking the albumin and
AFP genes in newborn and adult rat liver, adult kidney, the
7777-C8 hepatoma cell line and a rat cell line of fibroblastic
origin (65). Nuclease hypersensitive sites were probed with
DNase I by using the indirect end-labelling technique described by
Wu (63). Three DNase I-hypersensitive sites could be mapped in the
5'-end flanking region of the albumin gene in chromatin from
newborn and adult liver but not in the chromatin of the C8
hepatoma cell line and of non hepatic cells (adult kidney, JF-1
fibroblasts). Two of these sites are located within 0.5 Kb of the
5' end and the third site is found about 2.5 to 3 Kb upstream of
the albumin gene (65). The presence of these sites appears then to
be directly correlated with the actual state of transcription of
the albumin gene in the tissues and cell lines analysed.

Three distinct nuclease-hypersensitive sites have also been
found in the 5'-end flanking regions of the AFP gene in chromatin
from the actively transcribing newborn rat hepatocytes. All these
sites are also present in the C8 hepatoma cell line but could not
be detected in adult kidney or in the JF-1 fibroblasts cells.
Interestingly, only the more distal site located at about 2 to 3
Kb from the 5' end of the AFP gene is detected in adult liver
(Nahon and Sala-Trepat, in preparation). The presence of this site
could be related to the differenciated state of the cell while the
other two sites near the 5'-end of the AFP gene would be directly
correlated with active transcription. The fact that these two
sites are absent in adult liver provides the first indication for
the occurrence of alterations in chromatin structure that would
reflect the transition from the active to the inactive state of
the AFP gene during the terminal differenciation of the
hepatocyte.

CONCLUSIONS

The mammalian albumin and AFP genes provide a powerful model
system to investigate the molecular mechanisms responsible for
changes in gene expression during developmental and oncogenic
process. Different studies have indicated that the expression of
these two genes during rodent liver development and in different
hepatomas is regulated mainly at the transcriptional level.

Transcriptional control in eukaryotic cells can be achieved
by several possible mechanisms acting at the DNA genomic level or/
and at the chromatin level. Mechanism acting at the genomic level
could involve, for instance, rearrangement of genomic sequences,
changes in the relative location of the gene sequences within the
supercoiled loop domain or changes in the specific methylation of
cytosine residues.

We have analyzed the possible involvement of those mechanisms
in the modulation of the expression of the albumin and AFP genes
during rat liver development and neoplasia. Our studies have lead
to the following findings : 1) Changes in the transcriptional
template capacity of the albumin and AFP genes do not appear to
result from alterations in gene number or gross rearrangements
within the genome ; 2) Preferential association of the AFP gene
with the nuclear matrix is not a necessary requirement for its
active transcription in our cellular system ; 3) No evidence has
either been obtained for the implication of changes in methylation
of specific genes sequences in the transcriptional modulation of
these genes during rat liver development.

Our investigations on the chromatin structure of the AFP and
albumin genes in developing rat liver and other tissues and cell
lines have shown important differences in the conformation of
these genes in hepatic and non-hepatic cells. Further, different

sets of DNase I-hypersensitive sites have been found upstream from the albumin and AFP genes depending on the state of differentiation of the cells and on the transcriptional state of these genes in the tissues and cell lines analyzed. The particular chromatin structure at the 5' end of the albumin and AFP genes detected with the nuclease probe, is very likely to play an important role in the early establishment of the tissue-specific potential of overt gene expression and in the transcriptional regulation of these genes during the terminal differentiation. The short stretches of DNA amidst the 5' hypersensitive sites might mark sequences onto which regulatory proteins can be bound. It is now of obvious interest to search for specific proteins which will preferentially bind to these regions of the genome.

ACKNOWLEDGEMENTS

We thank Dr. T. Erdos and Dr. J.L. Danan for critical reading of the manuscript. This work was supported by grants from the Centre National de la Recherche Scientifique, the Institut National de la Santé et de la Recherche Médicale, and the Association pour le Développement de la Recherche sur le Cancer. A. Poliard was supported by a Postdoctoral Fellowship from the European Molecular Biology Organization, and M. Frain by a fellowship from the Ligue Nationale Française contre le Cancer.

REFERENCES

1) ABELEV, G.I. (1971). Alpha-fetoprotein in ontegenesis and its association with malignant tumors. Adv. Cancer Res. 14, 295-358.

2) SELL, S., BECKER, F., LEFFERT, H. and WATABE, H. (1976). Expression of an oncodevelopmental gene product (α-fetoprotein) during fetal development and adult oncogenesis. Cancer Res. 36, 4239-4249.

3) GITLIN, D. and BOESMAN, M. (1967). Sites of serum α-feto-protein synthesis in the human and in the rat. J. Clin. Invest. 46, 1010-1016.

4) RUOSLAHTI, E. and SEPPÄLÄ, M. (1979). α-Fetoprotein in cancer and fetal development. Adv. Cancer Res. 29, 275-346.

5) VAN FURTH, R. and ADINOLFI, M. (1969). In vitro synthesis of the fetal α_1-globulin in man. Nature 222, 1296-1299.

6) GITLIN, D. and GITLIN, J.D. (1975). Fetal and neonatal development of human plasma proteins. In "The Plasma Proteins" (F.W. Putnam, ed.). Vol. 2, pp. 263.319. Academic Press, New York.

7) SELLEM, C., FRAIN, M., ERDOS, T. and SALA-TREPAT, J.M. (1984). Differential expression of albumin and α-fetoprotein genes in fetal tissues of mouse and rat. Dev. Biol. 102, 51-60.

8) SELL, S., NICHOLS, M., BECKER, F.F. and LEFFERT, H.L. (1974). Hepatocyte proliferation and α_1-fetoprotein in pregnant, neonatal and partially hepatectomized rats. Cancer Research 34, 865-871.

9) WATABE, H. (1971). Early appearance of embryonic α-globulin in rat serum during carcinogenesis with 4-dimethylamino azobenzene. Cancer Res. 31, 1192-1194.

10) SELL, S., SALA-TREPAT, J.M., SARGENT, T., THOMAS, K., NAHON, J.L., GOODMAN, T.A. and BONNER, J. (1980). Molecular mechanisms of control of albumin and alpha-foetoprotein production: a system to study the early effects of chemical hepatocarcinogens. Cell Biol. Intern. Rep. 4, 235-254.

11) SCHREIBER, G., ROTERMUND, H.M., MAENO, H., WEIGAND, K. and LESHS, R. (1969). The proportion of the incorporation of leucine into albumin to that into total protein in rat liver and hepatoma Morris 5123TC. Eur. J. Biochem. 10, 355-361.

12) TSE, T.P.H., MORRIS, H.P. and TAYLOR, J.M. (1978). Molecular basis of reduced albumin synthesis in Morris hepatoma 7777. Biochemistry 17, 2121-2128.

13) SELL, S., THOMAS, K., MICHAELSON, M., SALA-TREPAT, J.M. and BONNER, J. (1979). Control of albumin and α-fetoprotein expression in rat liver and in some transplantable hepatocellular carcinomas. Biochim. Biophys. Acta 564, 173-178.

14) SALA-TREPAT, J.M., DEVER, J., SARGENT, T.D., THOMAS, K., SELL, S. and BONNER, J. (1979). Changes in expression of albumin and α-fetoprotein genes during rat liver development and neoplasia. Biochemistry 18, 2167-2178.

15) INNIS, M.A. and MILLER, D. (1977). Quantitation of rat α-fetoprotein messenger RNA with a complementary DNA probe. J. Biol. Chem. 252, 8469-8475.

16) MIURA, K., LAW, S.W., NISHI, S. and TAMAOKI, T. (1979). Isolation of α-fetoprotein messenger RNA from mouse yolk sac. J. Biol. Chem. 254, 5515-5521.

17) BROWN, P.D. and PAPACONSTANTINOU, J. (1979). Mouse albumin mRNA in liver and a hepatoma cell line. J. Biol. Chem. 254, 5177-5183.

18) SALA-TREPAT, J.M., SARGENT, T.D., SELL, S. and BONNER, J. (1979). α-Fetoprotein and albumin genes of rats : no evidence for amplification-deletion or rearrangement in rat liver carcinogenesis. Proc. Natl. Acad. Sci. USA 76, 695-699.

19) SARGENT, T.D., WU, J.R., SALA-TREPAT, J.M., WALLACE, R.B., REYES, A.A. and BONNER, J. (1979). The rat serum albumin gene : Analysis of cloned sequences. Proc. Natl. Acad. Sci. USA 18, 3256-3260.

20) JAGODZINSKI, L., SARGENT, T.D., YANG, M., GLACKIN, C. and BONNER, J. (1981). Sequence homology between RNAs encoding rat α-fetoprotein and rat serum albumin. Proc. Natl. Acad. Sci. USA 78, 3521-3525.

21) KIOUSSIS, D., EIFERMAN, F., RIJN, P.E., GORIN, M.B., INGRAM, R.J. and TILGHMAN, S.M. (1981). The evolution of α-fetoprotein and albumin. II. The structures of the α-fetoprotein and albumin genes in the mouse. J. Biol. Chem. 256, 1960-1967.

22) LAW, S., TAMAOKI, T., KREUZALER, M. and DUGAICZYK, A. (1980). Molecular cloning of DNA complementary to a mouse α-fetoprotein mRNA sequence. Gene 10, 53-61.

23) GAL, A., NAHON, J.L., LUCOTTE, G. and SALA-TREPAT, J.M. (1984). Structural variants of the α-fetoprotein gene in different inbred strains of rat. Mol. Gen. Genet. 195, 153-158.

24) SARGENT, T., JAGODZINSKI, L., YANG, M. and BONNER, J. (1981). Fine structure and evolution of the rat serum albumin gene. Mol. Cell. Biol. 1, 871-883.

25) EIFERMAN, F., YOUNG, P.R., SCOTT, R.W. and TILGHMAN, S.M. (1981). Intragenic amplification in the mouse α-fetoprotein. Nature 294, 713-718.

26) BROWN (1976) Structural origins of mammalian albumins. Fed. Proc. Am. Soc. Exp. Biol. 35, 2141-2144.

27) ALEXANDER, F., YOUNG, P.R. and TILGHMAN, S.H. (1984). Evolution of the albumin-α-fetoprotein ancestral gene from the amplification of a 27 nucleotide sequence. J. Mol. Biol. 173, 159-176.

28) INGRAM, R.S., SCOTT, R.W. and TILGHMAN, S.M. (1981). α-Fetoprotein and albumin genes are in tandem in the mouse genome. Proc. Natl. Acad. Sci. USA 78, 4694-4698.

29) SZPIRER, J., LEVAN, G., THORN, M. and SZPIRER, C. (1984). Gene mapping in the rat by mouse-rat somatic cell hybridization : Sinteny of the albumin and α-fetoprotein genes and assignment to chromosome 14. Cytogenet. Cell Genet. 38, 142-149.

30) DUGAICZYK, A., LAW, S.W. and DENNISON, D.E. (1982). Nucleotide sequence and the encoded amino-acids of human serum albumin mRNA. <u>Proc. Natl. Acad. Sci.</u> USA <u>79</u>, 71-75.

31) MORINAGA, T., SAKAI, M., WEGMANN, T.G. and TAMAOKI, T. (1983). Primary structures of human α-fetoprotein and its mRNA. <u>Proc. Natl. Acad. Sci. USA</u> <u>80</u>, 4604-4608.

32) FRAIN, M. (1984). Structure et expression des gènes codant pour deux protéines marqueurs de la différenciation hépatique chez l'homme : L'albumine et l'alpha-foetoprotéine. Thèse d'Etat, University of Paris.

33) HARPER, M.E. and DUGAICZYK, A. (1983) Linkage of the evolutionary related serum albumin and α-fetoprotein genes within q11-22 of human chromosome 4. <u>Am. J. Hum. Genet.</u> <u>35</u>, 565-572.

34) LIAO, W.S.L., CONN, A.R. and TAYLOR, J.M. (1980). Changes in rat α1-fetoprotein and albumin mRNA levels during fetal and neonatal development. <u>J. Biol. Chem.</u> <u>255</u>, 10036-10039.

35) CASSIO, D., WEISS, M.C., OTT, M.O., SALA-TREPAT, J.M., FRIES, J. and ERDOS, T. (1981). Expression of the albumin gene in rat hepatoma cells and their dedifferentiated variants. <u>Cell</u> <u>27</u>, 351-358.

36) BELANGER, L., FRAIN, M., BARIL, P., GINGRAS, M.C., BARTKOWIAK, J. and SALA-TREPAT, J.M. (1981). Glucocorticoid suppression of α_1-fetoprotein synthesis in developing rat liver. Evidence for selective gene repression at the transcriptional level. <u>Biochemistry</u> <u>20</u>, 6665-6671.

37) NAHON, J.L., GAL, A., FRAIN, M., SELL, S. and SALA- TREPAT, J.M. (1982). No evidence for post-transcriptional control of albumin and α-fetoprotein gene expression in developing rat liver and neoplasia. <u>Nucl. Acids Res.</u> <u>10</u>, 1895-1911.

38) TILGHMAN, S.H. and BELAYEW, A. (1982). Transcriptional control of the murine albumin/α-fetoprotein locus during development. <u>Proc. Natl. Acad. Sci. USA</u> <u>79</u>, 5254-5257.

39) GUERTIN, M., BARIL, P., BARTKOWIAK, J., ANDERSON, A. and BELANGER, L. (1983). Rapid suppression of αl-fetoprotein gene transcription by dexamethasone in developing rat liver. Biochemistry 22, 4296-4302.

40) DARNELL, J.E. (1982). Variety in the level of gene control in eukaryotic cells. Nature 297, 365-371.

41) WEISBROD, S. (1982). Active chromatin. Nature 297, 289-295.

42) TONEGAWA, S., SAKANO, H., MAKI, R., TRAUNECKER, A., HEINRICH, G., ROEDER, W. and KUROSAWA, Y. (1981). Somatic reorganisation in immunoglobin genes during lymphocyte differentiation. Cold Spring Harbor Symp. Quant. Biol. 45, 839-848.

43) KLEIN, G. (1983). Specific chromosomal translocations in the genesis of B-cell-derived tumors in mice and men. Cell 32, 311-315.

44) LUCOTTE, G., GAL, A., NAHON, J.L. and SALA-TREPAT, J.M. (1982). EcoRI restriction site polymorphism of the albumin gene in different inbred strains of rat. Biochem. Genetics 20, 1105-1115.

45) GAL, A., NAHON, J.L., LUCOTTE, G., ERDOS, T. and SALA-TREPAT, J.M. (1984). Structural basis for restriction site polymorphism at the albumin locus in inbred strains of rats. Biochem. Genetics, in press.

46) COOK, P.R. and BRAZELL, I.A. (1975). Supercoils in human DNA. J. Cell Sci. 19, 261-279.

47) BENYAHATI, C. and WORCEL, A. (1976). Isolation, characterization and structure of the folded interphase genome of *Drosophila melanogaster*, Cell 9, 393-407.

48) BEREZNEY, R. and COFFEY, D.J. (1974). Identification of a nuclear protein matrix. Biochem. Biophys. Res. Commun. 60, 1410-1417.

49) ROBINSON, S.I., NELKIN, B.D. and VOGELSTEIN, B. (1982). The ovalbumin gene is associated with the nuclear matrix of chicken oviduct cells. Cell 28, 99-106.

50) CIEJEK, E.M., NORDSTROM, J.L., TSAI, M.J. and O'MALLEY, B.W. (1982) Ribonucleic acid precursors are associated with the chick oviduct nuclear matrix. Biochemistry 21, 4945-4953.

51) COOK, P.R., LANG, J., HAYDAY, A., LANIA, L., FRIED, M., CHRISWELL, D.J. and WYKE, J.A. (1982). Active viral genes in transformed cells lie close to the nuclear cage. EMBO J. 1, 447-452.

52) CIEJEK, E.M., TSAI, M.J. and O'MALLEY, B.W. (1983). Actively transcribed genes are associated with the nuclear matrix. Nature 306, 607-609.

53) COOK, P.R. and BRAZELL, I.A. (1980). Mapping sequences in loops of nuclear DNA by their progressive detachment from the nuclear cage. Nucleic Acids Res. 8, 2895-2905.

54) VEDEL, M., GOMEZ-GARCIA, M., SALA, M. and SALA-TREPAT, J.M. (1983). Changes in methylation pattern of albumin and α-fetoprotein genes in developing rat liver and neoplasia. Nucl. Acids Res. 11, 4335-4354.

55) NAHON, J.L., GAL, A., ERDOS, T. and SALA-TREPAT, J.M. (1984). Differential DNase I sensitivity of the albumin and α-fetoprotein genes in chromatin from rat tissues and cell lines. Proc. Natl. Acad. Sci. USA 81, 5031-5035.

56) RAZIN, A. and RIGGS, A.D. (1980). DNA methylation and gene function. Science 210, 604-610.

57) TAYLOR, J.H. ed. (1983). DNA methylation and cellular differentiation. Cell Biology monographs. Vol. 11. Springer Verlag, Wien-New York.

58) NAKHASI, H.L., LUNCH, K.R., DOLAN, K.P., UNTERMAN, R.D. and FEIGELSON, P. (1981). Covalent modification and repressed transcription of a gene in hepatoma cells. Proc. Natl. Acad. Sci. USA 75, 834-837.

59) OTT, M.O., SPERLING, L., CASSIO, D., LEVILLIERS, J., SALA-TREPAT, J.M. and WEISS, M.C. (1982). Undermethylation at the 5'-end of the albumin gene is necessary but not

sufficient for albumin production by rat hepatoma cells in culture. Cell 30, 825- 833.

60) KUNNATH, L. and LOCKER, J. (1983). Developmental changes in the methylation of the rat albumin and α-fetoprotein genes. EMBO J. 2, 317-324.

61) WEINTRAUB, H. and GROUDINE, M. (1976). Chromosomal subunits in active genes have an altered conformation : Globin genes are digested by deoxyribonuclease I in red blood cell nuclei but not in fibroblast nuclei. Science 193, 848-856.

62) MATHIS, D., OUDET, P. and CHAMBON, P. (1980). Structure of transcribing chromatin. Progress Nucleic Acid Res. and Mol. Biol. 24, 1-54.

63) WU, C. (1980). The 5' end of Drosophila heat schock genes in chromatin are hypersensitive to DNase I. Nature 286, 854-860.

64) ELGIN, S.C.R. (1981). DNase I-hypersensitive sites of chromatin. Cell 27, 413-415.

65) NAHON, J.L. and SALA-TREPAT, J.M. (1984). Tissue-specific DNase I hypersensitive sites in rat chromatin are present upstream from the 5' ends of the albumin and α-fetoprotein genes. J. Cell Biol. 99, 139a.

TRANSCRIPTION CONTROL IN EUCARYOTES-ENHANCERS AND PROMOTERS

Brigitte Bourachot, Philippe Herbomel & Moshe Yaniv

Department of Molecular Biology
Pasteur Institute
25, rue du Docteur Roux
75015 Paris France

INTRODUCTION

The pioneering work of bacterial and phage geneticists have demonstrated that gene expression is regulated in response to changes in external conditions. The operon theory of Jacob & Monod (1) laid the basis for the identification of the transcription control sequences in procaryotes and of the proteins that interact with them. The conjunction of several approaches: genetic analysis, DNA cloning and sequencing, the isolation and characterization of control proteins and the development of powerful in vitro systems brought much insight to our present day understanding of the control of gene expression in procaryotes (see e.g. a recent review, (2,3)).

The parallel study of the mechanisms involved in the control of gene expression in eucaryotes was hampered by difficulties inherent to these systems; the lack of powerful genetic tools, the past difficulties in gene isolation and perhaps a more complex nature of the regulatory elements. Thus, recent developments in the isolation, sequencing and transfer of genes as well as the establishment of in vitro transcription systems in eucaryotes opened a

new era in this field. The comparison of the nucleotide sequences
preceding the cap site of many genes have revealed several common
features. The transcription start site (+1) is usually a purine
surrounded by pyrimidines. It is preceded 25 to 30 nucleotides to
its 5' side (position -25 to -30) by the TATA or Goldberg-Hogness
box showing the following sequence TATA^A_TA^A_T. Another consensus sequence GG^C_TCA^A_TCT or the CAAT box is found frequently further upstream at about -80 (4).

FUNCTIONAL STUDIES OF PROMOTERS

Studies aiming at determining the sequences upstream of the transcription initiation site that are required for the full activity
of a promoter were undertaken by several groups. For both the
herpes tk and the rabbit β-globin gene 110 nucleotide upstream of
the cap site are apparently sufficient to ensure full activity
after injection into xenopus oocytes or transfection of cells (5-
6). The situation became more complex when sequences controling
the early transcription of SV40 or polyoma viral DNA were studied.
It became apparent that nucleotides upstream of position -110 relative to the cap site (roughly between -113 to -275 including two
repeats of 72 b.p.) for SV40 (7-8) and (-170 to -450) for polyoma
(9-11) are required for full expression of the viral early promoter. Furthermore both in SV40 (7) and polyoma (9-11) the promoter proximal elements TATA or TATA plus CAAT boxes are less crucial for transcription than the far upstream sequences.

VIRAL ENHANCERS

Fromm & Berg (12) have shown that the 72 b.p. repeat element required for the transcription of the viral early functions can restore the viral viability when placed either inside the transcription unit (in the large T intron) or beyond the poly A addition
site. Independtly Banerji & Schaffner (13) have shown that the

72 b.p. element of SV40 can activate the rate of transcription of
the rabbit β-globin gene when placed either 5' or 3' to this gene
in both possible orientations. This *cis* acting DNA element was
called enhancer. Subsequent studies have shown that other DNA
tumor viruses or retroviruses contain enhancer elements either
preceding the cap site (polyoma, Adenovirus, Herpes, 14-18) or
following the polyA addition site (BPV, 19), or in both LTR elements of the integrated retroviruses (20).

TWO ENHANCERS COEXIST IN THE REGULATORY REGION OF POLYOMA

Polyoma virus multiplies in mouse fibroblasts or several other
differentiated cell lines but fails to grow on mouse early embryonal cells. Mutants of polyoma that overcome this block were isolated and analysed. It became apparent that all contain point mutations, duplications and deletion-duplications in the non coding
region of polyoma more precisely between the PvuII site at position 5267 and the BclI site at position 5032 (21). The same region
was shown by de Villiers & Schaffner (14) to contain the polyoma
enhancer. We wished to examine whether the enhancer function is
concentrated into a small DNA element -a binding site for a specific protein- or on the contrary is a property of an extended
DNA region. Our previous work on the structure of SV40 and polyoma
minichromosomes showing the existence of a DNaseI hypersensitive
domain along the 72 b.p. repeat element or the PvuII - BclI segment suggested that the second alternative may be closer to real-
ity (22-23).

To test the enhancer function of different DNA fragments we made
use of the chloramphenicol acetyl transferase (CAT) expression
vectors developed by Gorman *et al.* (24). As described in Fig. 1
we used CAT plasmids with the chicken α2 collagen promoter (either
a long or short version) and polylinkers either 5' or 3' to the
CAT transcription unit (15). Polyoma fragments were cloned either

in the 5' or the 3' sites. The different plasmids were transfected
into mouse fibroblasts by the calcium phosphate coprecipitation
procedure and cell extracts prepared and assayed for CAT activity
40 hours later. The enhancement factor was defined as the ratio
of CAT activities between enhancer containing and enhancer lacking
plasmids both containing the same promoter. To eliminate variations in transfection efficiencies between plate to plate we included in each transfection a reference plasmid - pSVE β-gal (pCH
110 of Hall *et al.* (25)). For each extract the activity of β-
galactosidase was measured and the CAT activities normalized to
constant β-gal activity.

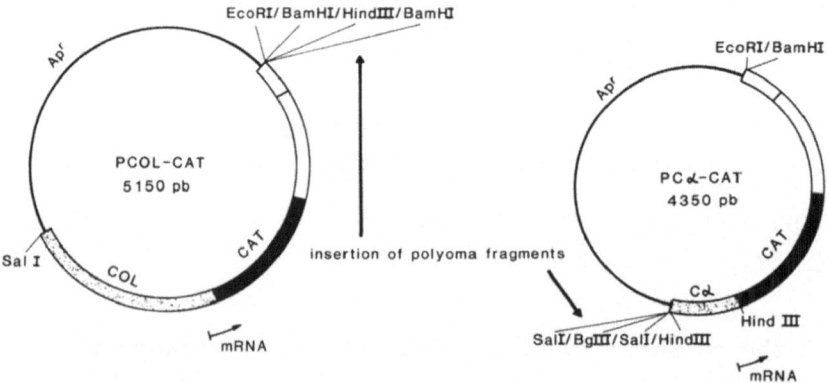

Fig. 1. *CAT expression vectors used to test the enhancer activity
of DNA fragments.* Polyoma DNA fragments were inserted in both
possible orientations at the Hind III site of pCOLCAT (3' position) or at the Bg1II site of pCαCAT (5' position). Thin lines are
pBR322 sequences; open boxes are SV40 sequences containing RNA
processing signals.

These normalized CAT activities were used in term to calculate the
enhancement factor. The results we obtained are summarized in Fig.
2. It is clear that each of the two neighbouring polyoma fragments
BclI-PvuII (element A) and PvuII-PvuII (element B) can enhance the
transcription from the collagen promoter when placed in either 3'
or 5' end. We checked that these fragments enhance initiations
from the correct start site by S1 mapping (15). Element A is a
strong enhancer in both orientations when placed at 5', but only
in one orientation at the 3'. ElementB shows a higher activity
when placed at 5' than when placed at 3'. Orientation dependent
activity of enhancers is observed frequently when short DNA fragments are used as enhancers (12,15). A longer DNA fragment containing both elements A and B functions in both orientations when
placed at 3' to the transcription unit (Fig. 2).

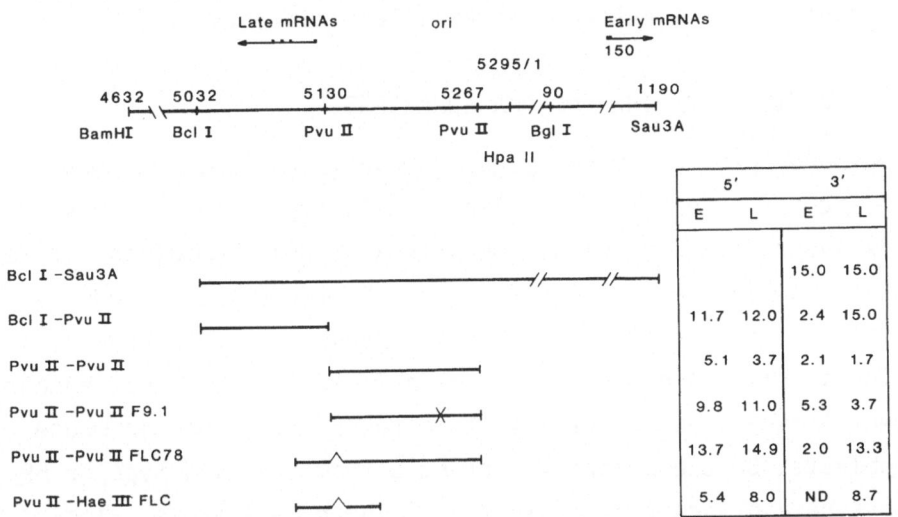

Fig. 2. *Enhancer activities of polyoma DNA fragments in mouse 3T6
fibroblasts.* The organization of the origin proximal sequences of
polyoma is shown at the top. The numbers given are the mean factor
of enhancement of CAT expression provided by insertion of the indicated polyoma fragment at 5' (2 left columns) and at 3' end of
the CαCAT or COLCAT transcription units, relative to the enhancer-
less plasmid. E and L mean that, respectively, the early and late
coding strand of the polyoma insert is in the same orientation as
the CAT coding strand. A cross on the PvuII-PvuII F9-1 fragment
shows the location of the F9-1 point mutation.

A point mutation at position 5233 -inside element B- isolated in
a polyoma mutant selected for growth in F9 mouse ambryonal carcinoma cells (pyECF9-1) increases the activity of this enhancer element by 2-3 fold. Trying to decrease the size of the enhancer
fragments that we cloned in our expression vectors failed to reveal any short sequence that maintains high activity. Comparing
different constructions suggest that the crucial sequences in element A are confined to 30 b.p. containing a core sequence
GCAGGAAG found also in the Adeno 5 Ela (16) and the mouse IgH enhancers (26-28). Element B contains the sequence TGTGGTTTTG homologous to the sequence common to SV40, mouse IgH, Ade 2 Ela and
MSV enhancers (29). The fact that short DNA fragments containing
these consensus sequences are inactive as enhancers suggest that
neighbouring nucleotides are not neutral and may contribute to
the activity of the enhancer as auxillary elements.

CELL SPECIFICITY OF THE POLYOMA ENHANCERS

As mentioned previously embryonal carcinoma cells of the mouse,
which are refractory to infection by wild type polyoma virus, are
permissive to certain polyoma mutants with point mutations or rearrangements in the Bc1I-PvuII region. We therefore tested the
activity of the different CAT constructions in PCC3 embryonal
carcinoma cells. The results given in Table 1 reveal that enhancer
element A is 3.5 fold less active in these cells than in mouse
fibroblasts. On the contrary element B from the wild type or F9
viruses retain identical activities in both cell types. Element B
from pyECF9-1 becomes now the major polyoma enhancer in mouse embryonal carcinoma cells.

Another example of an absolute cell specificity is shown by the
enhancer found between the variable and constant domains of the
rearranged heavy chain complex. This enhancer functions in lymph-

ocytes but not in mouse fibroblasts whereas the SV40 or polyoma
enhancers function in both cell types.

Table 1. Efficiencies of polyoma enhancer elements in 3T6 fibroblasts and PCC3 embryonal carcinoma cells

Fragment	Enhancer strength 5'E		Ratio 3T6/ PCC3	Enhancer strength 3'L		Ratio 3T6/ PCC3
	3T6	PCC3		3T6	PCC3	
Bcl I-Pvu II wt	11.7	3.6	3.3	15.0	4.3	3.5
Pvu II-Pvi II wt	5.1	3.0	1.7	1.7	2.1	0.8
Pvu II-Pvi II F9.1	9.8	9.6	1.0	3.7	3.9	0.9

The 5' data were obtained after insertion of enhancer fragments in
the BglIIsite of pCαCAT in the early orientation and in the Hind
IIIsite of pCOLCAT in the late orientation.

BIOLOGICAL ROLE OF ENHANCER ELEMENTS

The activation of the transcription unit of the immunoglobulin
heavy chains after the rearrangement of the V, D, J segments by
the IgH enhancer demonstrates the usefulness of such an element in
the biological context. The promoter or the enhancer are inactive
separately; only the rearrangement event turns on the transcription unit. In the case of DNA tumor viruses our recent experiments
suggest that the viral enhancers facilitate the transcription of
the early promoters in the presence of a large excess of cellular
transcription units competing for limiting transcription factors
or RNA polymerase (G. Moore & M. Yaniv, in preparation).

At least in the case of polyoma virus an enhancer is also required
for DNA replication. Either element A or B can supply the function

required for replication in the presence of T antigen complement-
ing in trans (30). Point mutations in the Ela type core sequence
of element A (in a virus lacking element B) abolish replication
(G. Magnusson, personal communication). Furthermore the replication and transcription function of the polyoma enhancer can be
replaced by the enhancer of SV40 or by the IgH enhancer. In this
last case the virus can replicate in mouse myeloma cells (31). In
this context we should mention the effect of the LTR elements of
retroviruses that can activate cellular genes by integration either
5' or 3' to a cellular gene (32).

ENHANCER SEQUENCES CREATE A UNIQUE CHROMATIN STRUCTURE

Both biochemical and electron microscopy techniques revealed the
existence of a unique structure on SV40 minichromosomes between
the origin of replication and the major start site of late RNA.
This region shows enhanced sensitivity to nucleases or restriction
enzymes (33-35) and appears as a nucleosome free segment of viral
DNA (22,36). The sensitivity of this region is not uniform. It
englobes several hypersensitive and resistant sites -e.g. the
three repetitions of the 21 b.p. (positions 40-103 in SV40 genome)
are totally resistant whereas three hypersensitive sites are found
around the origin and 4 along the two 72 b.p. repeats (37). Similarly the Bc1I-HpaII region of polyoma (positions 5022-5295) ex-
hibits an increased sensitivity to DNaseI with at least two hypersensitive sites in vicinity of the Ela and SV40 like enhancer core
sequences. Duplication of a 44 b.p. sequence containing the Ela
homology duplicates the DNaseI hypersensitive site (15,23).

SV40 recombinants with the enhancer placed at either the transcription termination site or inside the large T intron shows
DNaseI hypersensitivity (12) and a nucleosome free region in the
novel location of the enhancer sequences (38). It is interesting
to mention in this context that the immunoglobulin heavy or light

chains enhancers are also associated with a DNaseI hypersensitive site present only in IgH producing cells (39 and references within).

MECHANISM OF ACTION OF ENHANCERS

It is clear that enhancers increase the efficiency of existing promoters, however, in certain constructions they can also stimulate initiations from cryptic promoters (40). The effect of varying the distance between the enhancer and the target promoter on the transcriptional activity is controversial. In some cases the activity is distance dependent whereas in others it is not. In our experiments the action of element B is lower when placed at 3' to the gene whereas element A at least in one orientation shows equal activity at 3' and 5' sides of the gene (Fig. 2). In several cases the insertion of an active promoter between the enhancer and the target promoter reduces the final activity (41).

Taking all these facts into account favor the entry site model for enhancer function. The nucleosome free region serves as an entry site for transcription factors or RNA polymerase that will scan the DNA molecule bidirectionally. However, we cannot exclude for the moment that the enhancer can be a site of action for a topoisomerase that will propagate a perturbation in the DNA structure bidirectionnally. Another possible mechanism concerns an hypothetical anchoring to the nuclear matrix, the site of active transcription. It is clear that in vitro transcription studies should help decipher this intriguing observation (42). Since the enhancer can function even when placed in the middle of the transcription unit, it is reasonable to suggest that the enhancer activates the promoter by forming a stable initiation complex (43): it does not have to function continuously. In fact, recently myeloma variants were described were a deletion of the IgH enhancer does not impair the synthesis of immunoglobulins (44).

THE CONTRAST BETWEEN HIGHLY TISSUE SPECIFIC CELLULAR PROMOTERS AND
THE UBIQUITOUS VIRAL ENHANCER-PROMOTER SEQUENCES

The SV40 early promoter or the LTR of Rous sarcoma virus function
as strong promoters in a variety of cells from different species.
On the contrary, it was shown recently that the rat albumin promoter is only active in fully differentiated rat hepatocytes (45).
In these cells it is as active as the early promoter of SV40. On
the contrary it is inactive in dedifferentiated derivatives of
these cells, in albumin negative somatic cell hybrids between
hepatocytes and fibroblasts, in rat fibroblasts and even in mouse
hepatocytes. These experiments suggest that viral systems have
evolved to respond to a relatively large sepctrum of cells. They
can use transcription factors that are common to the majority of
mammalian cells. On the contrary the albumin promoter has to function only in a very specific cell. It may require a transcription
factor present only in a fully differenciated hepatic cell. Alternatively non hepatic cells may contain transcription repressor
that will block the expression of tissue specific promoters like
albumin. Another rat promoter coding for β-actin (46) ubiquitous
to all cells is active in several cell types of different species
(F. Thierry & M. Yaniv, in preparation). Other tissue specific
promoters were described recently for pancreatic cells by Walker
et al. (47). The insulin promoter functions only in endocrine
pancreatic cells whereas the chymotrypsin promoter functions only
in exocrine pancreatic cells. However, in these cases, the authors
did not observed species specificity at least between rat and man.
Other examples include the muscle specific rat α-actin promoter
that is expressed only after myoblast fusion (46) into myotubes or
the δ-crystallin gene that is more active in len cells (48). In
the case of insulin promoter, the upstream sequences can function
as a tissue specific enhancer (K. Yamamoto, personal communication)
when placed 3' to a ptk-CAT transcription unit. It is clear that
such results raise the question of the similitude and difference

between promoter upstream sequences and enhancer elements. For the
time being, we will like to keep the notion that enhancers function when placed 5' or 3' to a transcription unit, frequently in
all the four possible orientations, whereas promoter sequences
have to be close to the start site of transcription. Nevertheless,
these later can be sometimes turned around like the tk upstream
sequences or the distance between the different elements slightly
modified (49).

CONCLUSIONS

The complexity and large variety of control elements described
here suggest that a considerable number of transcription factors
can be found in eucaryotes. We may perhaps divide these factors
into two classes: (a) commitment proteins that bind to expression
control sequences during some step of the differentiation process
and maintain an open chromatin structure around the transcription
start site, (b) transcription factors that appear in the fully
differenciated cell or specific repressors that disappear at this
step, the combination of these elements will permit the efficient
transcription of a set of chosen genes. Observations of specific
DNaseI hypersensitive sites around the Drosophila heat shock gene
or the mouse α or β-globulin genes in mouse erythroleukemia cells
before their induction by heat or DMSO respectively supports this
hypothesis (50-51).

Faithful transcription of free DNA in vitro may either require
the second class of proteins only or both classes if the commitment factors are also part of the active transcription complex.
Both the DNA sequences that are involved in the specific transcription and in the establishment of the commitment step will be
required for the achievement of correct expression of genes in-
troduced in transgenic mice or flies. Enhancers somehow interfere
with such a scheme and can permit the efficient transcription of

viral or cellular promoters even if these cells lack some of the specific transcription elements.

REFERENCES

1) JACOB, F. & MONOD, J. (1961). Genetic regulatory mechanism in the synthesis of proteins. J. Mol. Biol., 3, 265.

2) DE CROMBRUGGHE, B., BUSBY, S., & BUC, H. (1984). Role of the cyclic AMP receptor protein in activation of transcription. Science, 224, 831.

3) RAIBAUD, O. & SCHWARTZ, M. (1984). Positive control of transcription initiation in bacteria. Annual Review of Genetics, 18, 173.

4) BREATHNACH, R. & CHAMBON, P. (1981). Organization and expression of eucaryotic split genes coding for proteins. Ann. Rev. Biochem., 50, 349.

5) MCKNIGHT, S.L. & KINGSBURY, R. (1982). Transcription control signals of a eukaryotic protein-coding gene. Science, 217, 316.

6) DIERKS, P., VAN OOYEN, A., COCHRAN, M.D., DOBKIN, C., REISER, J., & WEISSMANN, C. (1983). Three regions upstream from the cap site are required for efficient and accurate transcription of the rabbit β-globin gene in mouse 3T6 cells. Cell, 32, 695.

7) BENOIST, C. & CHAMBON, P. (1981). In vivo sequence requirements of the SV40 early promoter region. Nature, 290, 304.

8) GRUSS, P., DHAR, R., & KHOURY, G. (1981). Simian virus 40 tandem repeated sequences as an element of the early promoter. Proc. Natl. Acad. Sci. USA, 78, 943.

9) TYNDAL, C., LA MANTIA, G., THACKER, C.M., FAVALORO, J., & KAMEN, R. (1981). A region of the polyoma virus genome between the replication origin and late protein coding sequences is required in cis for both early gene expression and viral DNA replication. Nucl. Acids. Res., 9, 6231.

10) BENDIG, M.M., THOMAS, T., & FOLK, W.R. (1980). Regulatory mutants of polyoma virus defective in DNA replication and the synthesis of early proteins. Cell, 20, 410.

11) KATINKA, M. & YANIV, M. (1982). Deletions of N-terminal sequences of polyoma virus T-antigens reduce but do not abolish transformation of rat fibroblasts. Mol. Cell. Biol., 2, 1238.

12) FROMM, M. & BERG, P. (1983). SV40 early and late region promoter function are enhanced by the 72 base pair repeat inserted at distant locations and inverted orientations. Mol. Cell. Biol., 3, 991.

13) BANERJI, J., RUSCONI, S., & SCHAFFNER, W. (1981). Expression of a β-globin gene is enhanced by remote SV40 DNA sequences. Cell, 27, 299.

14) DE VILLERS, J. & SCHAFFNER, W. (1981). A small segment of polyoma-virus DNA enhances the expression of a cloned rabbit β-globin gene over a distance of at least 1400 base pairs. Nucl. Acids. Research, 47, 6251.

15) HERBOMEL, P., BOURACHOT, B., & YANIV, M. (1984). Two distinct enhancers with different cell specificities coexist in the regulatory region of polyoma. Cell, 39, 653.

16) HEARING, P. & SHENK, T. (1983). The adenovirus type 5 E1A transcriptional control region contains a duplicated enhancer element. Cell, 33, 695.

17) HEN, R., BORRELLI, E., SASSONE-CORSI, P., & CHAMBON, P. (1983). An enhancer element is located 340 base pairs upstream from the adenovirus-2 E1A cap site. Nucl. Acids. Res., 11, 8747.

18) LANG, J.C., SPANDIDOS, D.A., & WILKIE, N.M. (1984). Transcriptional regulation of a herpes simplex virus immediate early gene is mediated through an enhancer-type sequence. EMBO J., 3, 389.

19) LUSKY, M., BERG, L., WEIHER, H., & BOTCHAN, M. (1983). Bovine papillomavirus contains an activator of gene expression at the distal end of the early transcription unit. Mol. Cell. Biol., 3, 1108.

20) LEVINSON, B., KHOURY, G., VANDE WOUDE, G., & GRUSS, P. (1982). Activation of SV40 genome by 72-base pair tandem repeats of Moloney sarcoma virus. Nature, 295, 568.

21) KATINKA, M., VASSEUR, M., MONTREAU, N., YANIV, M., & BLANGY, D. (1981). Polyoma DNA sequences involved in control of viral gene expression in murine embryonal carcinoma cells. Nature, 290, 720.

22) SARAGOSTI, S., MOYNE, G., & YANIV, M. (1980). Absence of nucleosomes in a fraction of SV40 chromatin between the origin of replication and the region coding for the late leader RNA. Cell, 20, 65.

23) HERBOMEL, P., SARAGOSTI, S., BLANGY, D., & YANIV, M. (1981). Fine structure of origin-proximal DNAase I-hypersensitive region in wild type and EC mutant polyoma. Cell, 25, 651.

24) GORMAN, C.M., MOFFATT, L.F., & HOWARD, B.H. (1982). Recombinant genomes which express chloramphenicol acetyltransferase in mammalian cells. Mol. Cell. Biol., 2, 1044.

25) HALL, C., JACOB, E., RINGOLD, G., & LEE, F. (1983). Expression and regulation of *Escherichia Coli* lac Z gene fusions in mammalian cells. J. Mol. App. Genet., 2, 101.

26) BANERJI, J., OLSON, L., & SCHAFFNER, W. (1983). A lymphocyte specific cellular enhancer is located downstream of the joining region in immunoglobulin heavy chain genes. Cell, 33, 729.

27) GILLIES, S.D., MORRISON, S.L., OI, V.T., & TONEGAWA, S. (1983). A tissue specific transcription enhancer element is located in the major intron of a rearranged heavy chain gene. Cell, 33, 717.

28) NEUBERGER, M.S. (1983). Expression and regulation of immuno-
    globulin heavy chain gene transfected into lymphoid cells.
    EMBO J., 2, 1373.

29) WEIHER, H., KONIG, M., & GRUSS, P. (1983). Multiple point
    mutations affecting the simian virus 40 enhancer. Science,
    219, 626.

30) MULLER, M.J., MUELLER, C.R., MES, A., & HASSELL, J.A. (1983).
    Polyomavirus origin for DNA replication comprises multiple
    genetic elements. J. Virol., 47, 586.

31) DE VILLIERS, J., SCHAFFNER, W., TYNDALL, C., LUPTON, S., &
    KAMEN, R. (1984). Polyoma virus DNA replication requires an
    enhancer. Nature, 312, 242.

32) PAYNE, G.S., BISHOP, M.J., & VARMUS, H.E. (1982). Multiple
    arrangements of viral DNA and an activated host oncogene in
    bursal lymphomas. Nature, 295, 209.

33) SCOTT, W.A. & WIGMORE, D.J. (1978). Sites in SV40 chromatin
    which are preferentially cleaved by endonucleases. Cell, 15,
    1511.

34) VARSHAVSKY, A.J., SUNDIN, O.H., & BOHN, M.J. (1979). A stretch
    of "late" SV40 viral DNA about 400 b.p. long which included
    the origin of replication is specifically exposed in SV40
    minichromosomes. Cell, 16, 453.

35) WALDECK, W., FOHRING, B., CHOWDHURY, K., GRUSS, P., & SAUER,
    G. (1978). Origin of DNA replication in papovavirus chromatin
    is recognized by endogenous nuclease. Proc. Natl. Acad. Sci.
    USA, 77, 1068.

36) JAKOBOVITS, E.B., BRATOSIN, S., & ALONI, Y. (1980). A nu-
    cleosome free region in SV40 minichromosomes. Nature, 285,
    263.

37) CEREGHINI, S. & YANIV, M. (1984). Assembly of transfected
    DNA into chromatin: structural changes in the origin-promoter-
    enhancer region upon replication. EMBO J., 3, 1243.

38) JONGSTRA, J., REUDELHUBER, T.L., OUDET, P., BENOIST, C., CHAE, C.B., JELTSCH, J.M., MATHIS, D.J., & CHAMBON, P. (1984). Induction of altered chromatin structures by SV40 enhancer and promoter elements. Nature, 307, 708.

39) MILLS, F., FISHER, M., KURDA, R., FORD, A., & GOULD, H. (1984). DNase I hypersensitive sites in the chromatin of human μ immunoglobulin heavy chain genes. Nature, 306, 809.

40) WASYLYK, B., WASYLYK, C., AUGEREAU, P., & CHAMBON, P. (1983). The SV40 72 bp repeat preferentially potentiates transcription starting from proximal natural or substitute promoter elements. Cell, 32, 503.

41) DE VILLIERS, J., OLSON, L., BANERJI, J., & SCHAFFNER, W. (1983). Analysis of the transcriptional enhancer effect. In: Cold Spring Harbor Symposia on Quantitative Biology, Volume XLVII, p. 911, C.S.H.L.

42) SASSONE-CORSI, P., DOUGHERTY, J.P., WASYLYK, B., & CHAMBON, P. (1984). Stimulation in vitro transcription from heterologous promoters by the SV40 enhancer. Proc. Natl. Acad. Sci. USA, 81, 308.

43) BROWN, D.D. (1984). The role of stable complexes that repress and activate eucaryotic genes. Cell, 37, 359.

44) KLEIN, S., SABLITZKY, F., & RADBRUCH, A. (1984). Deletion of the IgH enhancer does not reduce immunoglobulin heavy chain production of a hybridoma IgD class switch variant. EMBO J., 3, 2473.

45) OTT, M.O., SPERLING, L., HERBOMEL, P., YANIV, M., & WEISS, M.C. (1984). Tissue specific expression is conferred by a sequence from the 5' end of the rat albumin gene. EMBO J., 3, 2505.

46) MELLOUL, D., ALONI, B., CALVO, J., YATTE, D., & NUDEL, U. (1984). Developmentally regulated expression of chimeric genes containing muscle actin DNA sequences in transfected myogenic cells. EMBO J., 3, 983.

47) WALKER, M.D., EDLUND, T., BOULET, A.M., & RUTTER, W.J. (1983).
    Cell specific expression controlled by the 5' flanking region
    of insulin and chymotrypsin genes. Nature, 306, 557.

48) KONDOH, H., YASUDA, K., & OKADA, T.S. (1983). Tissue specific
    expression of a cloned chick δ-crystallin gene in mouse cells.
    Nature, 301, 440.

49) MCKNIGHT, S.L., KINGSBURY, R.C., SPENCE, A., & SMITH, M. (1984).
    The distal transcription signals of the herpesvirus tk gene
    share a common hexanucleotide control sequence. Cell, 37, 253.

50) WU, C. (1984). Two protein-binding sites in chromatin implicat-
    ed in the activation of heat-shock genes. Nature, 309, 229.

51) HOFER, E., HOFER-WARBINEK, R., & DARNELL, J.E. (1982). Globin
    RNA transcription: A possible termination site and demonstra-
    tion of transcriptional control correlated with altered chroma-
    tin structure. Cell, 29, 887.

CONTROLS OF GENE EXPRESSION IN CHEMICAL CARCINOGENESIS: ROLE OF CYTOCHROME P450 MEDIATED MONO-OXYGENASES

M.C. Lechner

Laboratório de Bioquimica, Instituto Gulbenkian de Ciência, Apartado 14, 2781 Oeiras, Portugal

INTRODUCTION

After the first report by Sir Percivall Pott (1) in 1775, showing that scrotal cancer among chimney sweeps was due to occupational exposure to soot, many evidences have been accumulating in support of a major role played by the environment in the incidence of human cancers.

According to the statistics published by the World Health Organization already in 1964 (2), approximately 85% of all human cancers result directly or indirectly from environmental influences. Other reports reveal that 90% of all the environment related human cancers are in fact due to chemicals, and only the remaining 10% are attributable to viral, radiation and genetic factors (3,4). It is known nowadays that many environmental chemicals may be responsible for a considerable proportion of all human cancers, as the presence of chemical carcinogenic agents in our environment has been proved (5). They are derived from industrial products or by-products, polluted city air, drugs, certain smoked and cooked foods as well as from tobacco smoke, all together taking a great deal of importance in the ethiology of cancers (6).

Much interest exists in the identification and characterization of the genetic and environmental factors involved in chemical carcino-genesis as we are inevitably exposed to such agents by the ingestion of a constantly growing number of pollutants from different sources including pesticides, insecticides and other agricultural chemicals, contaminants of aquatic environment, marine and fresh water life, mainly from sewage treatment plants which affect the major part of the food chain. Also many occuptionally related chemicals can be a cause for the chemical aggression often resulting in a carcinogenic process.

Chemical Carcinogenesis is a Multistep Process

Chemical carcinogenesis is a highly complex multistep process de-pending upon the conjugation of exogenous and endogenous factors. The manifestation of a carcinogenic process depend on the genetic susceptibility of each individual as a genetic control can be exer-cised at virtually every stage of chemical carcinogenesis determin-ing its rate of progression.

The concept that carcinogenesis occur through different stages, depending on different kinds of stimuli and involving independent mechanisms, was first suggested by Friedwald and Rous (7), who proposed the terms of "Initiation" and "Promotion", based on the formation of skin tumors in rabbits by treatment with a carcino-genic agent - the initiator - and further action of nonspecific stimuli - the promoter. This theory has been confirmed to apply to all kinds of tissues (8-10), and the concepts of initiation and promotion further redefined in a more molecular and cellular basis.

As pointed out by Kouri *et al.* (11) three main stages can be dis-tinguished in a chemically induced carcinogenic process consisting in: Initiation, Biochemical Promotion and Cellular Promotion.

Stages in Chemical Carcinogenesis

The first step in chemical carcinogenesis depends on the exposure, uptake and distribution of the carcinogenic agent within the organism. The uptake is primarily determined by the level of exposure, however, the assimilation and distribution among the different organs and within the cells depend in many cases on the presence of specific receptors in the cytoplasm (12,13). There are important genetic differences in receptor concentration depending on the activity of the genes coding for these proteins.

Chemical carcinogens, like apolar xenobiotics in general, have to be transformed in order to cause their elimination from the organism. The metabolic pathways leading to the production of excretable polar end products are very complex and depend on several enzyme systems catalizing a sequence of reactions that constitute a biphasic process (14). The presence and relative concentrations of phase I and phase II enzymes is genetically determined. Cytochrome P450 mediated mono-oxygenases, responsible for most of the phase I reactions, acting on many exogenous compounds, are particularly important, as in some cases they can originate metabolites which are more cytotoxic and carcinogenic than the parent compound (15, 16).

The active agents or metabolites can be detoxified and eliminated from the cells and organisms, but they can instead bind to macromolecules, either without appreciable damage of the cells, or bind to specific macromolecules, in particular to DNA. The resulting DNA adducts can be eliminated by the cell repair systems or not, producing in this case genotoxic effects (17), namely initiation of a carcinogenic process. Promotion is considered to consist in the proliferation of a stable abnormal genotype originating transformation to the cancer phenotype (18).

Cellular promotion depends on both the quiescence of the cancer
cells and the immune system that can remove them from the body or
not. It is evident that at virtually every stage, the chemical
induced carcinogenic process is under genetic control as it depends
on the presence of specific proteins, receptors, enzymes or antibodies, products of gene expression of the host cell.

In this paper we will limitate the presentation to some aspects
concerning the involvement of the liver microsomal mono-oxygenases
both in the initiation and in the prevention of chemical carcinogenesis, and present data from our Laboratory on the molecular
mechanisms of action of a well known promoter of hepatic carcinogenesis, which is a potent inducer of the microsomal mono-oxygenases, the phenobarbital.

Initiation of Chemical Carcinogenesis and the Role of Mono-oxygenases

Initiators are genotoxic agents creating a memory effect (19). The
great majority of these chemicals are not direct carcinogens, but
are pre-carcinogens which must be activated through biotransformation reactions to reactive intermediates that are the proximate
or ultimate carcinogens, capable of directly initiate the process of
carcinogenesis (20). Compounds like aromatic amines and amides are
able to induce tumors in the liver and bladder, independently of
the way of administration, suggesting that they must be metabolized
in those tissues into carcinogenic active intermediates.

Ultimate carcinogens are strong electrophiles, reacting with nu-
cleophilic sites present in DNA molecules. It is known that the
modification of one base is enough to destabilize the double helix
at a several nucleotide distance, upstream and downstream the adduct. Apparently, only guanines have a decisive role in the initiation of chemical carcinogenesis (5), and a relationship has been
found between the modification of guanine, and genotoxic effects

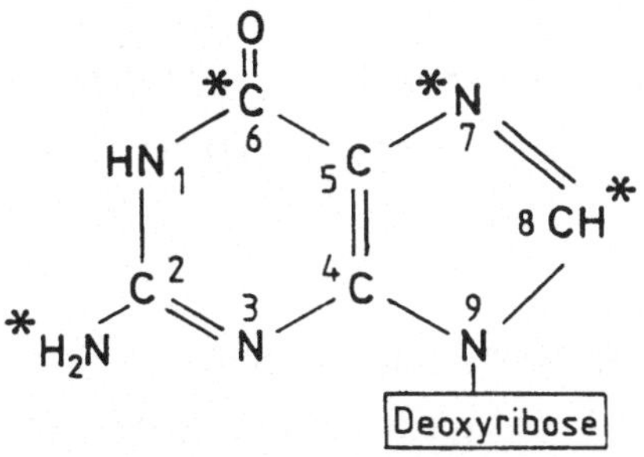

Figure 1.   *Sites for the formation of DNA adducts in guanine re-
            sidues, which have been related to Initiation of chem-
            ical carcinogenic processes by different agents.

of several carcinogenic agents which are represented in figure 1.
That is the case of BP, 7,8-diol-9,10 epoxide (21) and acetylaminofluorene (22) that bind to the guanine N_2 atom, aflatoxine, 2,3-
oxide that forms an adduct on N_7 (23), as well as the formation of
guanine adducts on C_8 or C_6 observed with acetylaminofluorene and
ethyl-nitrosourea, respectively, all these agents being known as
proximate carcinogens. Alkylation of O on C_6 of guanine residues
appears as the major determinant of carcinogenic effectiveness (14).
The irreversible nature of the initiators action seems to be due
to the genetic imprint produced by these genotoxic agents, that is
at the origin of subsequent tumor development.

One of the main mechanisms of bioactivation of pre-carcinogens is
the formation of arene-oxides by the liver microsomal mono-oxygenases - the aryl-hydrocarbon hydroxylases (AHH) - which are NADPH-
dependent cytochrome P450 mediated enzyme systems (24). Another
important mechanism for activation is the formation of nitrogen-
hydroxy metabolites. N-acetylamines, like 2 acetylaminofluorene are
transformed in the organism into carcinogenic intermediates through
nitrogen oxidation. Cytochrome P450 isozymes are responsible for
the metabolism of these substrates, polycyclic aromatic hydrocar-

bons, nitrosoguanidines and nitrosamines as well as many other xenobiotics like halogenated hydrocarbons, polyhalogenated biphenyls,
insecticides, amino-azodyes, aromatic amines and other heterocyclic
compounds (25).

P450 dependent enzyme systems can oxygenate PAH's at many positions
giving rise to different metabolites. The position specificity of
the different forms of cytochrome P450 plays a determinant role in
the activation or inactivation of chemical carcinogens (26), and
the existence of multiple forms of cytochrome P450's which are
integral membrane proteins of the endoplasmic reticulum explains
the broad substrate specificity of the hepatic microsomal mono-oxygenase system. These different isoenzymatic forms are inducible
at different extents by different chemical agents, playing a dual
role since in some cases they lead to partial or total detoxification, and in others to the production of active intermediates,
electrophyles capable of damaging critical cellular macromolecules.

PAH's that constitute a main group of environmental carcinogens,
are simultaneously substrates and potent inducers of the liver
microsomal cytochrome P450 dependent mono-oxygenases. Benz[a]pyrene
as well as methyl-cholantrene, prototypes of PAH's are metabolized
to carcinogenic intermediates, the 7,8-diol-9,10-epoxides predom-
inantly by P1-450 or P448 mediated mono-oxygenases. Actually, increased oxygenation of the PAH on non-K-region leading to diol-
epoxides formation is observed when there is a high P1-450/P450
ratio. Induction of this particular form of cytochrome P450 inevitably increases the generation of genotoxic metabolites and the incidence of tumorigenesis (25).

It has been demonstrated that P1-450 AHH production in the liver cells
is under the control of a particular genetic site, the *Ah locus*, which
plays a major role in the suceptibility to PAH tumorigenesis (27,28).

The *Ah locus* is proved to control the induction of the AAH's by
PAH's. The product of the Ah gene is a cytosolic receptor protein
capable of binding to PAH's, forming a complex, that is subsequently
translocated into the cell nucleus. There it interacts with DNA
producing a specific induction of cytochrome P1-450 dependent AHH's
(16). Activation of these structural genes by the receptor - PAH -
complex leads to increased synthesis of enzymes responsible for
the biotransformation of the inducer itself as well as other PAH's
into electrophylic intermediates, ultimate carcinogens. The susceptibility to PAH's-initiated tumors is indeed linked with inducible AHH activity.

The steady-state levels of direct carcinogens as well as ultimate
carcinogens, reactive electrophylic intermediates, are determinant
to the rates at which they interact with the critical nucleophylic
target (29). These steady-state levels depend on a delicate balance
between their detoxification and generation, both requiring the
intervention of different isoenzymatic forms of cytochrome P450
mono-oxygenases. Changes in that complex balance in specific tissues of an individual may therefore affect his risk of tumorigenesis.

Biochemical Promotion of Chemically Induced Carcinogenesis

Growth of initiated cells depend upon the application of appropriate growth stimuli. Tumor promoters act by increasing the proliferative activity of the island cells, improving the expression of
the tumor phenotype. It is believed that a common set of cellular
genes may help to mediate the genesis of the tumors (30) and that
promoters may act by stimulating an outstanding expression of those
genes, changing the cell homeostasis.

Promoters of Liver Carcinogenesis

Certain steroid hormones like progesterone, estradiol and mestranol, hypolipidemic agents, organochlorine pesticides as well as the anti-oxidant butylhydroxytoluene and phenobarbital have been recognized as liver carcinogenesis promoters (31). All of them are able to stimulate growth of the normal liver, causing marked hyperthrophy of the hepatocytes. In pre-initiated liver, they increase the proliferation of focal cells at pre-neoplastic state and accelerate islands enlargement, enhancing the manifestation of the proliferative advantage probably by increasing cell replication and delaying cell death or repair. In consequence, they increase the number of detectable islands on carcinogen-induced liver. These agents are incapable of inducing tumors when acting independently. They do not alter the degree of differentiation of the tumors, but prevent the reversion of altered foci that were induced by preceding carcinogen administration (32).

As stressed by Pitot and Sirica (18), very few studies have been directed towards an understanding of the mechanism of action of promoters effective in hepatocarcinogenesis. It is evident that the agents which have been demonstrated to act as promoters in the liver do not show enough analogies with the action of phorbol esters during skin carcinogenesis which have been more intensively studied (33). All the promoting agents that are active in liver chemical carcinogenesis are inducers of the mono-oxygenases and produce hypertrophy of the hepatocytes by stimulating an intense proliferation of the endoplasmic reticulum membranes.

Phenobarbital as a Promoter in Hepatic Chemical Carcinogenesis

Phenobarbital is the most widely used promoter in experimental hepatocarcinogenesis (18). When given to rats previously fed with 2-acetylaminofluorene, it promotes the development of differentiated hepatocarcinomas, initiated by 2-acetyl-aminofluorene, which

otherwise would not become manifest (34). The promoter effect of
phenobarbital is also evident when it is given to hepatectomized
diethylnitrosamine pre-treated rats, leading to the manifestation
of 6-8 times more islands than in the absence of phenobarbital (35).

Phenobarbital enhances the expression of the irreversible carcinogenic changes induced by genotoxic agents while by itself it does
not induce islands. The growth stimulus provided by phenobarbital
appears to be a determinant feature leading to stable alteration
and expression of the tumor phenotype.

Phenobarbital is the prototype of group I inducers of liver microsomal mono-oxygenases (36). When given to experimental animals, it
produces a pleiotypic response of the hepatic cell strongly stimulating the production of increased amounts of mono-oxygenases (37)
by inducing a particular form of cytochrome P450, the cytochrome
P450b, LM2 or 2, in the rat, rabbit or mouse, respectively (38,39).
This enzyme induction is concomitant with a marked proliferation
of the endoplasmic reticulum membranes resulting in a net liver
hypertrophy.

The mechanisms for the establishment of the liver phenobarbital
phenotype are complex and not yet clearly understood. Actually,
phenobarbital, in constrast to the benz[a]pyrene or 3-methyl-chlol-
anthrene group of liver mono-oxygenase inducers, does not compete
with the cytosolic receptor molecules (40) and there is no evidence
for any phenobarbital receptor protein present in hepatocytes.

Cytochrome P450 isoenzymatic form stimulated by phenobarbital is
structurally, genetically and enzymatically distinct from the one
that is stimulated by the polycyclic hydrocarbons, catalyzing the
hydroxylation of pre-carcinogens in different positions, which explains that it improves detoxication rather than conversion into

active intermediates like diol-epoxides (26,41). It is understand-
able that phenobarbital, although being a strong promoter of liver
chemical carcinogenesis when given after initiation by an active
carcinogen, when given to the experimental animals prior to the
administration of the carcinogen, confers a protection against
chemical carcinogenesis, reducing the binding of 2-acetylamino-
fluorene to DNA, by 80% (42,43).

Liver growth, but not induction of cytochrome P450 mono-oxygenases,
is considered to be the critical property of tumor promoters (31).

The elucidation of the molecular mechanisms by which phenobarbital
produces the growth stimulus, responsible for liver hypertrophy
and active expression of tumor phenotype in the initiated cells,
resulting in a further increase of the already enhanced proliferative activity of island cells is essential for a satisfactory
understanding of the carcinogenic process and its prevention and
control.

RESULTS AND DISCUSSION

Effects of Phenobarbital in Normal Rat Liver

We have been interested in the study of the course of biochemical
events concerning gene expression and regulation taking place during the onset of induction by phenobarbital in an attempt to find
a sequence of metabolic modifications produced by this agent, which
could contribute to the elucidation of the cascade modifications
that account for the modulation of protein synthesis, consequent
enzyme induction and endoplasmic reticulum membranes proliferation
brought about by this xenobiotic.

Phenobarbital affects the protein pattern of the liver cell by
selectively inducing the synthesis of some endoplasmic reticulum

components without affecting the amino acid incorporation into the proteins of other sub-cellular compartments (44,45). A single dose of phenobarbital elevates in 3-4 hours, more than twofold the synthesis of nascent peptides on membrane-bound polysomes (46), while protein synthesis by free polysomes does not present significant changes under PB action (47,48). The accumulation of active mRNA engaged in the protein synthesis by bound polysomes is observed shortly after one single phenobarbital administration, without any evidence for an increase of the total pool of intranuclear pre-mRNA and mRNA metabolism (49). The half-life of RNA is increased in the liver by pre-treatment with phenobarbital (50) and inversely related to the activity of microsomal alkaline ribonuclease (51) which is associated with proliferation and important reorganization of the endoplasmic reticulum membranes.

Our studies also showed that the synthesis of membrane proteins induced by phenobarbital in the rat liver is associated to a marked increase in poly(A)$^+$ mRNA in the liver microsomes (52). RNA labelling kinetics *in vivo* studied by the administration of [^{14}C] orotic acid, as well as the determination of the template activities of microsomal RNA's (53) give evidence of the accumulation of active mRNA associated with the endoplasmic reticulum membranes, without a corresponding increase in the incorporation of the labelled precursor into total nuclear RNA (54). Consistent with this observation is the absence of stimulation of nuclear RNA polymerases I, II and III over a period between 3 hours and 4 days after phenobarbital administration (55), and the fact that any discernible increase in [^{32}P] incorporation into total nuclear, nucleolar and nucleoplasmic RNA can be detected during induction by phenobarbital (56). The activity of the nuclear poly(A) polymerase did not reveal any detectable increase after phenobarbital administration, compatible with the accumulation of poly(A)$^+$ mRNA in the microsomes (57).

Gene expression in eukaryotic cells is regulated to a large extent at post-transcriptional levels. The existence of precursors to mRNA's containing by far more RNA sequences than the functional mRNA's, and particularly the discontinuous form in which the sequences are present in DNA precludes the existence of complex regulatory mechanisms for the intra-nuclear processing of nuclear-cytoplasmic transport for each specific messenger (58).

Messenger RNA exists in animal cells as ribonucleoprotein complexes in different pools. Informosomes, the mRNP particles in the cytoplasm of secretory cells exist not only in polyribosomal, free and bound mRNP's, but also as free short-term and long-term repressed mRNA, absent from the polyribosomal mRNA populations. A kinetic relationship between non-polysomal "silent" messenger RNA and polyribosomal "active" mRNA, consistent with a precursor-product relationship between the respective mRNP's, has been found in different biological systems. Many different mRNA species, coding for specific proteins have been identified in the post-ribosomal supernatant of a variety of tissues.

The existence of these potentially functional mRNA's stocked in the cytoplasm in a latent state as mRNP complexes, constitute an important device for translational control mechanisms, as there is a reversible equilibrium between polysomes and free mRNP's plus ribosomal sub-units. This is a particularly important device for the control of protein synthesis in embryonic systems. A pool of poly(A)$^+$ free mRNP particles exists in rat liver (60), to which a biological role as a precursor of active polyribosomal membrane bound messenger is ascribed. Post-ribosomal supernatant of normal rat liver contains 15% of the total poly(A)$^+$ mRNA present in the cytoplasm, and a large part of the total ferritin mRNA (44%) is found in the post-ribosomal supernatant of normal rat liver. A cytoplasmic control mechanism of iron stimulated ferritin synthesis has been proved to exist in the liver (61) where iron treatment causes a dramatic decrease in the post-ribosomal ferritin mRNA, and a corresponding increase in the polyribosomal ferritin pool. Similarly, the pool of albumin mRNA present in the liver post-ribosomal supernatant increases dramatically in a short-term fast attaining up to 60% of total cytoplasmic albumin mRNA sequences. Albumin mRNA can be stored as cytoplasmic free mRNP during the fasting state as the reduced rate of albumin synthesis in fasting can be rapidly reversed by feeding or by the supplementation of the animals or cells with amino acids, through a rapid reassembly of active polysomes (62).

A two-fold increase in the rate of *in vivo* cytochrome P450 apoprotein synthesis is observed in the liver 4-6 hours after the administration of phenobarbital (63), while the lag period for detecting increased amounts of *in vitro* translatable mRNA in the total poly(A)$^+$ mRNA has been demonstrated to be of the order of 16 hours (64).

We have previously observed that total poly(A)$^+$ mRNA isolated from rat liver, 24 hours after a single phenobarbital administration, is more active in stimulating amino acids incorporation into several proteins, including cytochrome P450 inducible apoprotein, when assayed *in vitro* in a reticulocyte lysate (65). Concomitantly, poly(A)$^+$ RNA in the free cytoplasmic RNP particles decreases to 68% of the value found before phenobarbital administration, while the poly(A)$^+$ RNA associated to the endoplasmic reticulum membranes rises to 157% of the normal value (54).

Labelling kinetics studies of the poly(A)$^+$ RNA in the free cytoplasmic RNP particles revealed the presence of large amounts of *de novo* synthesized messengers, suggesting that at least part of these free informosome pool consists of potentially active phenobarbital inducible messengers coding for endoplasmic reticulum proteins.

The relative amount of poly(A) sequences in the total poly(A)$^+$ RNA
both from bound polysomes and f.c RNP's is significantly higher in
rat liver after phenobarbital administration (54). This can account
for the increased template activity previously found, as it is
known that 3'OH poly(A) tails play a fundamental role in the biogenesis of mRNA and its utilization in translation by eukaryotic
cells, the length of these structures being related to the stability and translation efficiency of the messengers (66).

The intracytoplasmic mRNA distribution and particularly the free
RNP pool in the liver is markedly affected after administration of
phenobarbital, suggesting that mobilization of potentially active
messages can be an early event contributing for the regulation of
gene expression mechanisms displayed during the action of this
chemical agent.

Spacial segregation of mRNA's to be translated on ribosomes bound
to the endoplasmic reticulum is essential in the biogenesis of
secretory as well as endoplasmic reticulum membrane proteins (67-
69), and a direct association of mRNA and membranes can contribute
to facilitate their utilization (70,71). It has been demonstrated
that cytochrome P450 mRNA in phenobarbital treated rat liver is
primarily associated with ribosomes bound to the endoplasmic reticulum membranes, the cytochrome P450 apoprotein being exclusively
synthesized by polysomes associated with the endoplasmic reticulum,
and directly inserted into the membranes (72). We have shown that
an important enhancement in the capacity of stripped microsomes
from induced livers to bind RNP complexes is produced at very
early stages (2-4 hrs) of the adaptative response to phenobarbital
(73), compatible with the hypothesis that a primary effect of
phenobarbital in the liver must be a movement of stocked mRNA from

f.c.RNP's into active bound-polysomes (74).

Quantification of total poly(A)$^+$ mRNA populations have been performed together with the investigation of specific mRNA's coding
for inducible (cytochrome P450b) and non-inducible (albumin) proteins, which are synthesized by bound polysomes, by immunoprecipitation of the *in vitro* translated polypeptides and specific cDNA
hybridization (75). Our results point at a selective mobilization
of stored mRNA's with segregation of mRNA classes for translation
in association with the endoplasmic reticulum membranes to occur at
an early stage of the adaptative response of the liver cell to this
xenobiotic. Modulation of translation by membranes is a potentially
powerful way to influence gene expression in the hepatic cell. However, recent studies performed with specific molecular probes of
labelled cDNA corresponding to the mRNA's of NADPH-cytochrome c
oxido reductase and cytochrome P450b, proteins that are induced by
phenobarbital treatment, indicate that there is an increase in
nuclear as well as in cytoplasmic sequences corresponding to these
mRNA's which is produced early after the administration of this
xenobiotic (76,77), although these increases are masked when total
populations are studied as described above.

In vitro recombinant DNA technology is a powerful and unique approach to analyse changes and modulation of gene expression particularly in somatic cells.

Hybridization of transformed colonies from a library of cDNA sequences with probes prepared from populations of poly(A)$^+$ RNA
from a tissue in different developmental or physiological stages
allows the identification of clones containing messenger sequences
differently represented in the RNA preparations. This methodology
also gives a semi-quantitative analysis of the changes in abundance
of specific sequences in the RNA preparations, and finally it al-

lows the selection of cloned sequences presenting a characteristic
type of expression in its isolation and further utilization as
probes to the analysis of the molecular mechanisms responsible for
the changes in expression of the respective genes.

We have applied this methodology to the study of the modulation of
gene expression in rat liver by PB. We have constructed a library
of recombinant bacterial plasmids containing cDNA copies of polysomal poly(A)$^+$ RNA obtained from the livers of phenobarbital pre-
treated animals (80 mg/kg body weight, 16 hours before sacrifice),
in order to analyse differential gene expression induced by this
xenobiotic, by comparing it with an identically prepared pBR322
library of cDNA sequences from untreated rat liver.

A thousand cloned sequences of each library have been screened by
double-cross colony hybridization with [^{32}P] cDNA prepared from
the same, homologous and heterologous poly(A)$^+$ RNA's, and the
distribution of the cloned sequences according to their abundance
has been determined. The comparative analysis revealed the existence of many sequence groups corresponding to mRNA's which are
present in different concentrations in rat liver after phenobarbital treatment (78), showing that this chemical agent, in parallel
to the induction of increased translation of a significant number
of active mRNA sequences, represses the translation of an equally
significant number of polysomal mRNA sequences.

In order to investigate the mechanisms involved in regulating the
expression of the phenobarbital modulated sequences, induced and
repressed, we selected from the screened libraries, clones containing cDNA inserts corresponding to mRNA's showing relative abundance
changes under phenobarbital treatment. After amplification, the
recombinant pDNA's were used as probes for dot hybridization and
northern blotting analysis of total and polysomal poly(A)$^+$ RNA at
different induction stages.

The fact that during the onset of phenobarbital response, the changes in the concentration of total mRNA does not always parallel the evolution in active polysomal mRNA in the same way for each of the mRNA sequences studied, demonstrates that modulation of protein synthesis by phenobarbital is brought about by complex mechanisms involving both transcriptional and translational regulation.

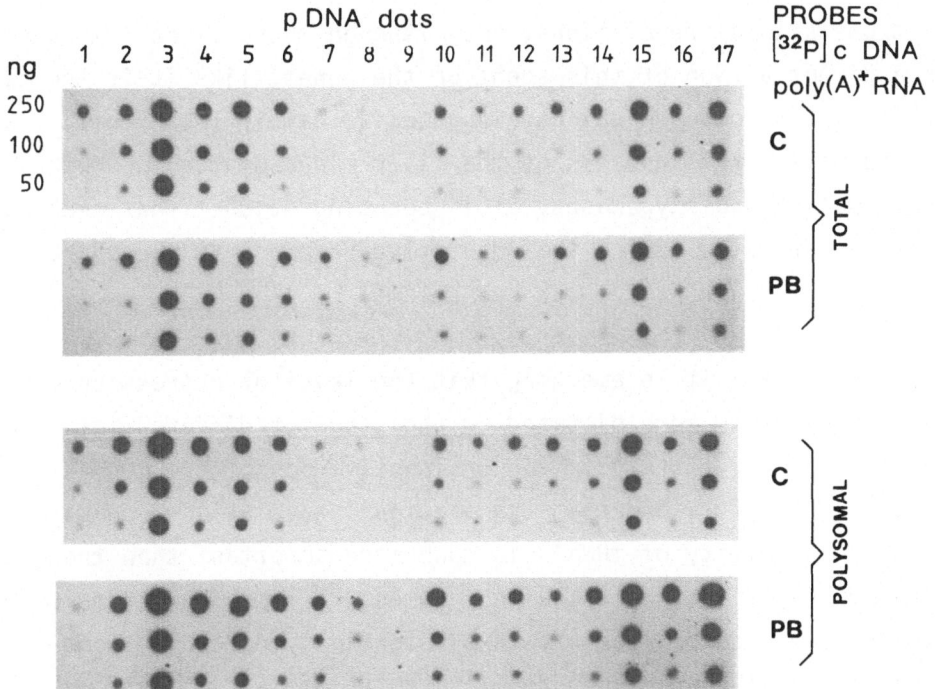

Figure 2. *Hybridization of control and phenobarbital [^{32}P] cDNA
           probes from total and polysomal poly(A)$^{+}$ RNA to 250, 100
           and 50 ng dots of pDNA from 17 selected recombinant
           plasmids.* Repression of sequence represented in clone 1
           is observed, affecting polysomal mRNA population more
           markedly than the total. Clones 7, 8, 16 and 17 are
           representative of inductions produced by phenobarbital,
           affecting total (clone 7), or preferentially polysomal
           active mRNA (clones 8, 16 and 17).

CONCLUSION

The ability of phenobarbital to promote the development of liver
tumors from previously initiated cells is much probably related to
the liver growth response induced by this xenobiotic, which in the
non-initiated cells result in a net hypertrophy. Proliferation of
the endoplasmic reticulum membranes produced at the early stages
of phenobarbital action on the liver cells reflects a change in the
liver cell homeostasis which responds to the chemical aggression
by changing the pattern of protein synthesis. The available data
prove that stimulation of liver growth by phenobarbital is the
result of very complex mechanisms involving both transcriptional
and translational regulations, and although there is no evidence
for a direct action of this agent on the genes, like it is the case
for initiators of chemical carcinogenesis, namely those that, like
phenobarbital, are inducers of the liver microsomal cytochrome P450
isoenzymatic mono-oxygenases, it is becoming evident that gene activity is affected during the adaptative response to phenobarbital,
possibly by unknown feed-back mechanisms.

From recent data it is emerging that the level of expression of
liver proteins is more dictated by the stage of differentiation of
the cell than by the activation or inactivation of the gene itself
(Deschatrette, J.), and that translational regulation mechanisms,
including stability of mRNA's is much more important than thought
up to now (Sperling, L.). The enhancement of common proteins characteristic of adult liver growth substantially depend on post-
transcriptional controls (Darnell, J.E.) (79).

There are important modifications occurring very early under phenobarbital on the distribution and utilization of mRNA's within the
cytoplasm of the liver cell, associated with a reorganization of
the endoplasmic reticulum membranes, the target of tumor promoters.
These phenomena, more than a direct action of specific genes ex-

pression, may be determinant and constitute the trigger for the growth response, which in initiated cells results in the promotion of chemical carcinogenesis processes.

Considering the hypothesis that the process of carcinogenesis can be the result of an outstanding expression of an otherwise normal gene which can be the fatal flow for cancerous growth (80) and that a common set of cellular genes can help to mediate the genesis of all tumors, phenobarbital could act not as a specific tumor promoter, but simply by changing the homeostasis of the cells, modulating the expression of a multitude of genes mainly through epigenetic mechanisms, improving expression of the tumor phenotype.

ACKNOWLEDGEMENTS

Original research described in this article was performed with the collaboration of C. Sinogas, M.L. Osório-Almeida, M.T. Freire and the technical assistance of Aline A. Bettencourt. We would like to thank J.-M. Sala-Trepat and his co-workers of the Laboratoire d'Enzymologie, CNRS, Gif-sur-Yvette, France, for the valuable collaboration in the cloning work.

REFERENCES

1) POTT, P. (1775). Cancer scrotic, in Chirurgical Observations, Hawes, Clarke and Collins, London, 63.

2) World Health Organization, Prevention of Cancer, Tech. Rep. Ser. No. 276, World Health Organization, Geneva (1964).

3) HIGGINSON, J. (1969). Present trends in cancer epidemiology, Proc. Can. Cancer Res. Conf., 8, 40.

4) BOYLAND, E. (1967). The correlation of experimental carcinogenesis and cancer in man, Progr. Exp. Tumor Res., 11, 222.

5) DAUNE, M. & FUCHS, R.P.P. (1980). La cancerogenese chimique, La Recherche, 115, 1066.

6) SCHECHTMAN, L.M., HENRY, C.J. & KOURI, R.E. (1980). Exposure,
    uptake and distribution of chemical carcinogens, in Genetic
    Differences in Chemical Carcinogenesis, R.E. Kouri, ed., CRC
    Press, Inc.

7) FRIEDWALD, W.F. & ROUS, P. (1944). The initiation and promot-
    ing elements in tumor production, J. Exp. Med., 80, 101.

8) ARMUTH, V. & BERENBLUM, I. (1972). Systemic promoting action
    of phorbol in lung and liver carcinogenesis in AKR mice,
    Cancer Res., 32, 2259.

9) ARMUTH, V. & BERENBLUM, I. (1974). Promotion of mammary car-
    cinogenesis and leukemogenic action by phorbol in virgin
    female Wistar rats, Cancer Res., 34, 2704.

10) PERAINO, C., FRY, R.J.M., STAFFELDT, E. & CHRISTOPHER, J.P.
    (1975). Comparative enhancing effects of phenobarbital, amino-
    barbital, diphenylhydantoin and dichlorodiphenyltrichloro-
    ethane on 2-acetyl-aminofluorene-induced hepatic tumorigene-
    sis in the rat, Cancer Res., 35, 2884.

11) KOURI, R.E., HENRY, C.J. & KREISHER, J.H. (1980). Stages in
    carcinogenesis, in Genetic Differences in Chemical Carcino-
    genesis, CRC Press, Inc., Boca Raton, Florida, R.E. Kouri, ed.

12) NEBERT, D.W., ROBINSON, J.R., NIWA, A., KUMAKI, K. & POLAND,
    A.P. (1975). Genetic expression of aryl hydrocarbon hydroxy-
    lase activity in the mouse, J. Cell Physiol., 83, 393.

13) POLAND, A.P., GLOVER, E. & KENDE, A.S. (1976). Stereospecific
    high affinity binding of 2,3,7,8, tetrachlorodibenzo-p-dioxin
    by hepatic cytosol: evidence that the binding species is the
    receptor for the induction of aryl-hydrocarbon hydroxylase,
    J. Biol. Chem., 251, 4936.

14) WILLIAMS, R.T. (1959). Detoxification mechanisms, in The
    Metabolism and Detoxification of Drugs, Toxic Substances and
    other Organic Compounds, 2nd ed. John Wiley & Sons, New York.

15) GELBOIN, H.V., HUBERMAN, E. & SACHS, L. (1969). Enzymatic
    hydroxylation of benzopyrene and its relationship to cyto-

toxicity, Proc. Natl. Acad. Sci. USA, 64, 1188.

16) NEBERT, D.W. (1981). Genetic differences in susceptibility to chemically induced myelotoxicity and leukemia, Environm. Health Perspectives, 39, 11.

17) MARQUARDT, H.W.J. (1980). DNA-The critical cellular target in chemical carcinogenesis?, in Chemical Carcinogens and DNA, Vol. II, Ch. VI, ed. P.L. Grover, CRC Press.

18) PITOT, H.C. & SIRICA, A.E. (1980). The stages of initiation and promotion in hepatocarcinogenesis, B.B.A., 605, 191.

19) EMMELOT, P. & SCHERER, E. (1980). The first relevant cell stage in rat liver carcinogenesis. A quantitative approach, B.B.A., 605, 247.

20) MILLER, E.C. (1978). Some current perspectives on chemical carcinogenesis in humans and experimental animals, Cancer Res., 38, 1479.

21) WEINSTEIN, I.B., JEFFREY, A.M., JENNETTE, R.W., BLOBSTEIN, S. H., HARVEY, R.G., HARRIS, C., AUTRUP, H., KASAI, H. & NAKANISHI, K. (1976). Benzo[a]pyrene diol epoxides as intermediates in nucleic acid binding *in vitro* and *in vivo*, Science, 193, 592.

22) FUCHS, R.P.P., LEFEVRE, J.-F., POUYET, J. & DAUNE, M.P. (1976). Comparative orientation of the fluorene residue in native DNA modified by N-acetoxy-N-2 acetylaminofluorene and two 7-halogene derivatives, Biochemistry, 15, 3347.

23) ESSIGMANN, J.M., CROY, R.G., NADZAN, A.M., BUSBY JR., W.F., REINHOLD, V.N., BUCHI, G. & WOGAN, G.N. (1977). Structural identification of the major DNA adduct formed by aflatoxin B1 *in vitro*, Proc. Natl. Acad. Sci. USA, 74, 1870.

24) GELBOIN, H.V., KINOSHITA, N. & WIEBEL, F.J. (1972). Microsomal hydroxylases: induction and role in polycyclin hydrocarbon carcinogenesis and toxicity, Fed. Proc., Fed. Am. Soc. Exp. Biol., 31, 1298.

25) COON, M.J. & VATSIS, K.P. (1978). Biochemical studies on chem-
    ical carcinogenesis: role of multiple forms of liver micro-
    somal cytochrome P450 in the metabolism of benzo[a]pyrene and
    other foreign compounds, Polycyclic Hydrocarbons and Cancer, ed.
    Academic Press, Inc., 1, 335.

26) NEBERT, D.W. & JENSEN, N.M. (1979). The *Ah locus*: genetic re-
    gulation of the metabolism of carcinogens, drugs and other
    environmental chemicals by cytochrome P450 mediated mono-oxy-
    genases, CRC Critical Reviews in Biochemistry, 401.

27) POLAND, A.P., GLOVER, E., ROBINSON, J.R. & NEBERT, D.W. (1974).
    Genetic expression of aryl hydrocarbon hydroxylase activity.
    Induction of mono-oxygenase activities and cytochrome P1-450
    formation by 2,3,7,8-tetrachlorodibenzo-p-dioxin in mice genet-
    ically "non-responsive" to other aromatic hydrocarbons, J.
    Biol. Chem., 249, 5599.

28) NEBERT, D.W. (1981). Genetic differences in susceptibility to
    chemically induced myelotoxicity and leukemia, Environm.
    Health Perspect., 39, 11.

29) BERENBLUM, I. (1975). Sequencial aspects of chemical carcino-
    genesis, Skin Cancer, F.F. Becker, ed., Plenum Press, New
    York, 1, 323.

30) BISHOP, J.M. (1983). Cancer genes come to age, Cell, 32, 1018.

31) SCHULTE-HERMANN, R., SCHUPPLER, R.J., OHDE, G., BURSCH, W. &
    TIMMERMANN-TROSIENER, I. (1982). Phenobarbital and other liver
    tumor promoters, in Chemical Carcinogenesis, C. Nicolini, ed.,
    Plenum Press, New York and London.

32) WATANABE, K. & WILLIAMS, G.M. (1978). Enhancement of rat hepa-
    tocellular-altered foci by the liver tumor promoter pheno-
    barbital: evidence that foci are precursors of neoplasms and
    that the promoter acts on carcinogen-induced lesions, J. Natl.
    Cancer Inst., 61, 1311.

33) O'BRIAN, P.J., SIMSIMAN, R.C. & BOUTWELL, R.K. (1975). Induct-
    ion of the polyamine biosynthetic enzymes in mouse epidermis

by tumor promoting agents, Cancer Res., 35, 1662.

34) PUGH, T.D. & GOLDFARB, S. (1978). Quantitative histochemical and autoradiographic studies of hepatocarcinogenesis in rats fed 2-acetyl-aminofluorene followed by phenobarbital, Cancer Res., 38, 4450.

35) PITOT, H.C., BARSNESS, L., GOLDSWORTHY, T. & KITAGAWA, T. (1978). Biochemical characterisation of stages of hepatocarcinogenesis after a single dose of diethylnitrosamine, Nature, 271, 456.

36) MANNERING, G.J. (1968). Significance of stimulation and inhibition of drug metabolism in pharmacological testing, Selected Pharmacological Testing Methods, ed. M. Burger Dekker, N.Y., 51.

37) ORRENIUS, S. & ERNSTER, L. (1964). Phenobarbital induced synthesis of the oxidative demethylating enzyme of rat liver microsomes, Biochim. Biophys. Res. Commun., 16, 60.

38) RYAN, D.E., THOMAS, P.E., KORZENIOWSKI, D. & LEVIN, W. (1979). Separation and characterisation of highly purified forms of liver microsomal cytochrome P450 from rats treated with polychlorinated biphenyls, phenobarbital and 3-methylchlolanthrene, J. Biol. Chem., 254, 1365.

39) HAUGEN, D.A. & COON, M.J. (1976). Properties of electrophoretically homogeneous phenobarbital-inducible and β-napthoflavone-inducible forms of liver microsomal cytochrome P450, J. Biol. Chem., 251, 7929.

40) POLAND, A. & GLOVER, E. (1975). Genetic expression of arylhydrocarbon hydroxylase by 2,3,7,8-tetrachlorodibenzo-p-dioxin: evidence for a receptor mutation in genetically non-responsive mice, Mol. Pharmacol., 11, 389.

41) THORGEIRSSON, S.S. & NEBERT, D.W. (1977). The Ah locus and the metabolism of chemical carcinogens and other foreign compounds, Adv. Cancer Res., 25, 149.

42) PERAINO, C., FRY, R.J.M., STAFFELD, E. & KISIELESKI, W.E.
(1973). Effects of varying the exposure to phenobarbital on
its enhancement of 2-acetylaminofluorene-induced hepatic
tumorigenesis in the rat, Cancer Res., 33, 2701.

43) PERAINO, C., FRY, R.J.M., STAFFELDT, E. & CHRISTOPHER, J.P.
(1977). Enhancing effects of phenobartitone and butylated
hydroxytoluene on 2-acetylaminofluorene-induced hepatic tumorigenesis in the rat, Food Cosmet. Toxicol., 15, 93,

44) KATO, R., LOEB, L. & GELBOIN, H.V. (1965). Microsome-specific
stimulation by phenobarbital of amino acid incorporation *in
vivo*, Biochem. Pharmacol., 14, 1164.

45) ARIAS, I.M., DOYLE, A. & SCHIMKE, R.T. (1969). Studies on the
synthesis and degradation of protein of the endoplasmic recti-
culum of rat liver, J. Biol. Chem., 244, 3303.

46) GLAZER, R.I. & SARTORELLI, A.C. (1972). The effect of phenobarbital on the synthesis of nascent protein on free and membrane-bound polyribosomes of normal and regenerating liver,
Molec. Pharmacol., 8, 701.

47) McCAULEY, R. & COURI, D. (1971). Early effects of phenobarbital
on cytoplasmic RNA in rat liver, Biochim. Biophys. Acta, 238.

48) FREIRE, M.T. & LECHNER, M.C. (1985). Manuscript in preparation.

49) LECHNER, M.C., SINOGAS, C.M., FREIRE, M.T. & BRAZ, J. (1982).
Expression of liver mono-oxygenase functions induced by xeno-
biotics, in Somatic Cell Genetics, C.T. Caskey, ed., Plenum
Publishing Corporation, 69.

50) STEELE, W.J. (1970). Phenobarbital induced prolongation of the
half-life of ribosomal-RNA of rat liver, Fed. Proc. Fed. Am.
Societies Exp. Biol., 29, 737.

51) LECHNER, M.C. & POUSADA, C.R. (1971). A possible role of liver
microsomal alkaline ribonuclease in the stimulation of oxidative drug metabolism by phenobarbital, chlordane and chloro-
phenothane (DDT), Biochem. Pharmacol., 20, 3021.

52) LECHNER, M.C. (1976). Effect of phenobarbital treatment on poly(A)-rich RNA in rat liver microsomes, I.U.B. Xth Internat. Congress Biochem. Hamburg, 03-6-130.

53) LECHNER, M.C. (1974). Studies of RNA from rat liver endoplasmic reticulum sub-fractions. Effect of phenobarbital treatment, Naunyn-Schmiedeberg's Arch. Pharm., Supp. 285, R50.

54) LECHNER, M.C. & SINOGAS, C.M. (1980). Changes in gene expression during liver microsomal enzyme induction by phenobarbital Biochem. Biophys. and Regulation of Cytochrome P450, Gustafsson *et al.*, ed., Elsevier/North-Holland, 405.

55) LINDRELL, T.J., ELLINGER, R., WARREN, J.T., SUNDHEIMER, D. & O'MALLEY, A.F. (1977). The effect of acute and chronic phenobarbital treatment of the activity of rat liver DNA dependent RNA polymerases, Molec. Pharm., 13, 426.

56) KUMAR, A.,SATYANARAYANA RAO, R. & PADMANABAN, G. (1980). A comparative study on the early effects of phenobarbital and 3 methylcholanthrene on the synthesis and transport of ribonucleic acid in rat liver, Biochem. J., 186, 81.

57) LECHNER, M.C. & SINOGAS, C.M. (1978). Studies on liver poly(A) rich RNA during microsomal enzyme induction, 12th FEBS Meeting, Dresden, RDA, 1157, 124.

58) SCHERRER, K., IMAIZUMI-SCHERRER, M.T., REYNAUD, C.A. & THERWATH, A. (1979). On pre-messenger RNA and transcription. A review, Mol. Biol. Rep., 5, 5.

59) MAUNDRELL, K., MAXWELL, E.S., CIVELLI, O., VINCENT, A., GOLDENBERG, S., BURI, J.F., IMAIZUMI-SCHERRER, M.T. & SHERRER, K. (1979). Messenger RNP complexes in avian erithroblasts: carriers of post-transcriptional regulation?, Mol. Biol. Rep., 5, 43.

60) HEMMINKI, K. (1975). Labelling kinetics of RNA containing poly(A) in liver sub-cellular fractions, Molec. and Cell Biochem., 8, 123.

61) ZÄHRINGER, J., BALIGA, B.S. & MUNRO, H.N. (1976). Novel mechanism for translation control in regulation of ferritin synthesis by iron, Proc. Natl. Acad. Sci., 73, 857.

62) YAP, S.H., STRAIR, R.K. & SHAFRITZ, D.A. (1978). Effect of a short term fast on the distribution of cytoplasmic albumin messenger ribonucleic acid in rat liver, J. Biol. Chem., 253, 4944.

63) BHAT, K.S. & PADMANABAN, G. (1978). Cytochrome P450 synthesis in vivo and in a cell-free system from rat liver, FEBS Lett., 89, 337.

64) DUBOIS, R.N. & WATERMAN, M.R. (1979). Effect of phenobarbital administration to rats on the level of the in vitro synthesis of cytochrome P450 directed by total rat liver RNA, Biochem. Biophys. Res. Commun., 90, 150.

65) LECHNER, M.C., FREIRE, M.T. & GRONER, B. (1979). In vitro biosynthesis of liver cytochrome P450 mature peptide sub-unit by translation of isolated poly(A)$^+$ mRNA from normal and phenobarbital induced rats, Biochem. Biophys. Res. Commun., 90 531.

66) NOKIN, P., HUEZ, G., MARBAIX, G., BURNY, A. & CHANTRENNE, H. (1976). Molecular modifications associated with aging of globin messenger RNA in vivo, Eur. J. Biochem., 62, 509.

67) SHIRES, T.K. & PITOT, H.C. (1974). The membron: a functional hypothesis for the translational regulation of genetic expression, Biomembranes, L.A. Manson, ed., Plenum Press, New York, 5, 81.

68) RICHTER, J.D. & SMITH, L.D. (1981). Differential capacity for translation and lack of competition between mRNA's that segregate to free and membrane-bound polysomes, Cell, 27, 183.

69) AMAR-COSTESEC, A., TODD, J.A., SABATINI, D.D. & KREIBICH, G. Characterization of the translocation apparatus of the endoplasmic reticulum. I-Functional tests of rat liver microsomal subfractions, personal communication.

70) CARDELI, J., LONG, B. & PITOT, H.C. (1976). Direct association of messenger RNA labelled in the presence of fluoro-orotate with membranes of the endoplasmic reticulum in rat liver, J. Cell Biol., 70, 47.

71) LANE, M.A., ADESNIK, M., SUMIDA, M., TASHIRO, Y. & SABATINI, D.D. (1975). Direct association of messenger RNA with microsomal membranes in human diploid fibroblasts, J. Cell Biol., 65, 513.

72) BAR-NUN, S., KREIBICH, G., ADESNIK, M., ALTERMAN, L., NEGISHI, M. & SABITINI, D.D. (1980). Synthesis and insertion of cytochrome P450 into endoplasmic reticulum membranes, Proc. Natl. Acad. Sci., 77, 965.

73) SINOGAS, C.M. & LECHNER, M.C., manuscript in preparation.

74) LECHNER, M.C. & SINOGAS, C.M. (1981). The importance of RNP's/membrane interactions for stimulation of protein synthesis by phenobarbital, Biochem. Soc. Transact., 9, 156 P.

75) FREIRE, M.T. & LECHNER, M.C., manuscript in preparation.

76) GONZALEZ, F.J. & KASPER, C.B. (1982). Cloning of DNA complementary to rat liver NADPH-cytochrome c (P450) oxidoreductase and cytochrome P450b mRNA's, J. Biol. Chem., 257, 5962.

77) HARDWICK, J.P., GONZALEZ, F.J. & KASPER, C.B. (1983). Transcriptional regulation of rat liver epoxide hydrase, NADPH-cytochrome P450 oxidoreductase and cytochrome P450b genes by phenobarbital, J. Biol. Chem., 258, 8081.

78) LECHNER, M.C., SINOGAS, C.M., OSÓRIO-ALMEIDA, M.L., CHAUMET-RIFFAUT, PH. & SALA-TREPAT, J.M. Phenobarbital mediated modulation of gene expression in rat liver: analysis of rat liver cDNA clones, manuscript in prepration.

79) INSERM Conference, Molecular biology and pathology of hepatic differentiation, Seillac, France, 21-26 October 1984.

CONTRIBUTORS

BOURACHOT, B., Department of Molecular Biology, Pasteur Institute, 25, rue du Docteur Roux, 75015 Paris, France. p. *267*

CELIS, A., Division of Biostructural Chemistry, Department of Chemistry, Aarhus University, DK-8000 Aarhus C, Denmark. p. *223*

CELIS, J.E., Division of Biostructural Chemistry, Department of Chemistry, Aarhus University, DK-8000 Aarhus C, Denmark. p. *223*

CIRILLO, D., Institute of Histology, University of Torino, Medical School, C.so M. D'Azeglio 52, 10126 Torino, Italy. p. *97*

COMOGLIO, P.M., Institute of Histology, University of Torino, Medical School, C.so M. D'Azeglio 52, 10126 Torino, Italy. p. *97*

CROCE, C.M., The Wistar Institute of Anatomy and Biology, Philadelphia, PA 19104, USA. p. *65*

CUZIN, F., Unité de Génétique Moléculaire des Papovavirus (INSERM U273), Centre de Biochimie, Université de Nice, Parc Valrose, 06034, Nice France. p. *127*

DI RENZO, M.F., Institute of Histology, University of Torino, Medical School, C.so M. D'Azeglio 52, 10126 Torino, Italy. p. *97*

DUESBERG, P.H., Department of Molecular Biology, University of California, Berkeley, California 94720, USA. p. *21*

FERGUSON, P., Department of Molecular Genetics, Smith Kline and French Laboratories, Philadelphia, Pennsylvania, USA. p. *167*

FERRACINI, R., Institute of Histology, University of Torino, Medical School, C.so M. D'Azeglio 52, 10126 Torino, Italy. p. *97*

FRAIN, M., Laboratoire d'Enzymologie, C.N.R.S., 91190 Gif-sur-Yvette, France. p. *239*

FRANZA, B.R., Cold Spring Harbor Laboratory, Cold Spring Harbor, N.Y. 11724, USA. p. *209*

GAL, A., Laboratoire d'Enzymologie, C.N.R.S., 91190 Gif-sur-Yvette, France. p. *239*

GARRELS, J.I., Cold Spring Harbor Laboratory, Cold Spring Harbor, N.Y. 11724, USA. p. *209*

GIANCOTTI, F.G., Institute of Histology, University of Torino, Medical School, C.so M. D'Azeglio 52, 10126 Torino, Italy. p. *97*

GIORDANO, S., Institute of Histology, University of Torino, Medical School, C.so M. D'Azeglio 52, 10126 Torino, Italy. p. *97*

GOMEZ-GARCIA, M., Laboratoire d'Enzymologie, C.N.R.S., 91190 Gif-
    sur-Yvette, France. p. *239*
GRAESSMANN, A., Institut fuer Molekularbiologie und Biochemie,
    Freie Universitaet Berlin, D-1000 Berlin 33, FRG. p. *113*
GRAESSMANN, M., Institut fuer Molekularbiologie und Biochemie,
    Freie Universitaet Berlin, D-1000 Berlin 33, FRG. p. *113*
GRIFFIN, B.E., Department of Virology, Royal Postgraduate Medical
    School, Hammersmith Hospital, London W12, England. p. *135* and
    157
HERBOMEL, P., Department of Molecular Biology, Pasteur Institute,
    25, rue du Docteur Roux, 75015 Paris, France. p. *267*
HIRT, B., Swiss Institute for Experimental Cancer Research, Ch.
    des Boveresses, 1066 Epalinges, Switzerland. p. *175*
HUNTER, T., Molecular Biology and Virology Laboratory, The Salk
    Institute, Post Office Box 85800, San Diego, California 92138,
    USA. p. *79*
JONES, N., Purdue University, West Lafayette, Indiana, USA. p. *167*
KAN, N., Laboratory of Molecular Oncology, National Cancer Insti-
    tute, Frederick Cancer Research Facility, Frederick, Maryland
    21701, USA. p. *21*
KRIPPL, B., Laboratory of Molecular Genetics, National Institute
    of Child Health and Human Development, National Institutes of
    Health, Bethesda, Maryland, USA. p. *167*
LECHNER, M.C., Laboratório de Bioquímica, Instituto Gulbenkian
    de Ciencia, Oeiras, Portugal. p. *285*
MARCHISIO, P.C., Institute of Histology, University of Torino,
    Medical School, C.so M. D'Azeglio 52, 10126 Torino, Italy.
    p. *97*
NAHON, J.L., Laboratoire d'Enzymologie, C.N.R.S., 91190 Gif-sur-
    Yvette, France. p. *239*
NALDINI, L., Institute of Histology, University of Torino, Medical
    School, C.so M. D'Azeglio 52, 10126 Torino, Italy. p. *97*
NOWELL, P.C., Department of Pathology and Laboratory Medicine,
    University of Pennsylvania School of Medicine, Philadelphia,
    PA 19104, USA. p. *65*
NUNN, M., The Salk Institute, P.O. Box 85800, San Diego, Califor-
    nia 92138-9216, USA. p. *21*
PAPAS, T., Laboratory of Molecular Oncology, National Cancer In-
    stitute, Frederick Cancer Research Facility, Frederick, Mary-
    land 21701, USA. p. *21*
POLIARD, A., Laboratoire d'Enzymologie, C.N.R.S., 91190 Gif-sur-
    Yvette, France. p. *239*
RASSOULZADEGAN, M., Unité de Génétique Moléculaire des Papovirus
    (INSERM U273), Centre de Biochimie, Université de Nice, Parc
    Valrose, 06034, Nice, France. p. *127*
ROSENBERG, M., Laboratory of Molecular Genetics, National Institute
    of Child Health and Human Development, National Institutes of
    Health, Bethesda, Maryland, USA. p. *167*

SALA-TREPAT, J.M., Laboratoire d'Emzymologie, C.N.R.S., 91190 Gif-sur-Yvette, France. p. *239*

SARIN, P.S., Laboratory of Tumor Cell Biology, National Cancer Institute, Bethesda, Maryland 20205, USA. p. *185*

SEEBURG, P.H., Genentech, Inc., 460 Point San Bruno Boulevard, South San Francisco, California 90007, USA. p. *21*

SIVAK, A., Biomedical Research and Technology Section, Arthur D. Little, Inc., Acorn Park, Cambridge, Massachusetts 02140, USA. p. *1*

STREULI, C., Imperial Cancer Research Fund, Lincoln's Inn Fields, London WC2, England. p. *135*

TARONE, G., Institute of Histology, University of Torino, Medical School, C.so M. D'Azeglio 52, 10126 Torino, Italy. p. *97*

TRATNER, I., Laboratoire d'Enzymologie, C.N.R.S., 91190 Gif-sur-Yvette, France. p. *237*

TU, A.S., Biomedical Research and Technology Section, Arthur D. Little, Inc., Acorn Park, Cambridge, Massachusetts 02140, USA. p. *1*

WATSON, D., Laboratory of Molecular Oncology, National Cancer Institute, Frederick Cancer Research Facility, Frederick, Maryland 21701, USA. p. *21*

WESTPHAL, H., Laboratory of Molecular Genetics, National Institute of Child Health and Human Development, National Institutes of Health, Bethesda, Maryland, USA. p. *167*

YANIV, M., Department of Molecular Biology, Pasteur Institute, 25, rue du Docteur Roux, 75015 Paris, France. p. *267*

INDEX

Adult T cell leukemia/lymphoma
    (ATLL), 187
AGMK cells, 159
Ah locus, 290
AIDS, 185
Albumin, 239
    locus, 239
    promoter, 276
Abelson murine leukemia virus
    (AMuLV), 97
Activated proto-*onc* genes, 21
Adeno-associated viruses (AAV),
    179
Adeno fiber protein, 115
Adenoviruses, 130, 167
    E1A, 130, 167
    E1B, 130
Adhesion plaques, 105
Allelic polymorphism, 247
Alpha-fetoprotein, 239
Alpha-fetoprotein locus, 239
AMA, 225
Amplification, 244
Antiviral drugs, 199
ATLV, 189
Avian carcinoma viruses, 25
    CMII, 25
    MH2, 25
    OK10, 25
Avian erythroblastosis virus
    (E26), 26
Avian myeloblastosis virus
    (AMV), 27
Avian sarcoma virus (ASV), 97

β-actin, 276
BALB/c-3T3, 2
B-cell growth factor, 194
B-cell lineage, 68
B-cell lymphomas, 32
B-lymphocytes, 157
Burkitt's lymphoma, 34, 65, 159

Cancer, 22, 65, 135, 175
    carcinomas of mice, 39
    human B-cell tumors, 34, 65
    human bladder carcinoma, 38
    human melanomas, 41
    initiation, 22
    maintenance genes, 22
    mammary carcinomas of rat, 39
    mouse plasmacytoma, 36
    papova viruses, 135
    parvo viruses, 175
    promotion, 22
    sinergestic, 45
    single-gene determinants, 23
Carcinogens, 1, 289
    chemical, 285
    identification, 1
    obligatory, 24
Carcinoma of mice, 39
Carcinoma virus MC29, 22
Cell cycle, 4, 223
Cell division, 209, 223
Cell DNA synthesis, 114, 226
Cell line, 30
Cellular proliferation, 163, 223
C3H-10T½, 2
Chemical carcinogenesis, 285
    role of mono-oxygenases, 288
    stages, 287

Chemotherapy, 199
Chloramphenicol acetyltransferase
    (CAT), 269
Chromatin structure, 255
Chromosomes
    2, 66
    8, 66
    11, 73
    14, 66
    18, 73
    22, 66
Chromosome translocation, 34, 66
Chronic leukemia viruses, 22
Clonal assay, 2
Complementation test, 168
cRNA injection, 114
Cosmid library, 158
Cotransfections, 146
Cyclin, 210, 212, 223
    cell cycle, 222
    cell division, 223
    distribution, 231
    DNA replication, 223
    immunofluorescence localiza-
        tion, 225
    S-phase, 227
Cytochrome P450, 289
    apoprotein, 297
    isozymes, 289
Cytoplasmic free mRNP, 297

Dividing cells, 176
DNA, 181
    methylation, 250
    microinjection, 113
    rearrangements, 244
    replication, 223
    SV40, 113
    synthesis, 226
DNAse I hypersensitive mini-
    chromosomes, 269
DNAse I hypersensitive sites,
    257, 274
DNAse I sensitivity, 256
Downstream promotion, 32
Drugs, 199

E1A, 130
E1A gene, 167

E1A proteins, 167
    functional domains, 167
    microinjection, 168
EBNA, 162
E1B, 130
EGF, 83
EGF receptor, 82
ELISA assays, 191
Embryonal cells, 269
Enhancers, 268, 269
Enhancer elements, 69
Entry site model, 275
Episomal mutant, 148
Epithelial cells, 157
    immortalization, 157
Epstein-Barr virus (EBV), 157

Feline sarcoma virus (FeSV), 97
Fisher rat embryo-Raucher
    leukemia virus, 5
Focus assay, 5
Fujinami sarcoma viruses, 24, 97

G1/S border, 227
Gamma interferon, 194
Gene expression, 239, 267
Glycolytic enzymes, 83
Growth at low serum concentration
    129
Growth regulation, 209
GTPase, 80

Hairy cell leukemia, 188
Harvey sarcoma virus, 22, 30
HeLa protein catalogue, 223
Hepatocytes, 252, 276
Human bladder carcinoma, 38
Human foetal kidney, 161
Human melanomas, 41
Human T cell leukemia virus
    (HTLV), 185
Hydroxyurea, 231
HTLV
    antibodies, 190
    transmission studies, 191
HTLV-1, 188
HTLV-2, 188
HTLV-3, 188

Immortalising genes, 157
Immortalization, 40, 157
Immonoglobulin heavy chain locus, 66
Immunoglobulin light chain loci, 69
Informosomes, 296
Initiation, 22
In situ hybridization, 253
Integration, 181
Interleukin, 3, 194

Kirsten sarcoma viruses, 29

Leukemias, 65
Leukemogenesis, 194
Liver
    albumin, 239
    alpha-feto protein, 239
    carcinogenesis, 240, 292
Long-range enhancers, 69
Lymphokines, 193

Maintenance genes, 22
Mammary carcinomas of rats, 39
Marmoset kidney cells, 161
Membrane proteins
    synthesis, 295
Metabolic activation, 8
Microinjection, 168
    cRNA, 114
    DNA, 113
    EIA protein, 168
Mitogens, 88
Moloney sarcoma virus, 30
Monoclonal antibodies, 145
Mono-oxygenases, 285
Mouse plasmacytoma, 36
mRNA, 295
Mycosis fungoides, 186
Multigene-one cancer hypothesis, 21

Neoplastic transformation, 1, 21, 65, 79, 113, 127, 135, 157, 167, 175, 209, 239
    advantages and disadvantages, 1
    adenovirus EIA gene, 167
    altered morphology, 11

anchorage independent growth, 12
assays, 2
Epstein-Barr virus, 157
immortality, 12, 40, 157
immunoglobulin genes, 65
oncogenes, 13, 21, 65, 79
oncogenicity *in vivo*, 11
papova viruses, 135
parvoviruses, 175
phosphoproteins, 97
polyoma virus, 127
protein data bases, 209
proto-*onc* genes, 21
SV40 DNA fragments, 113
systems, 1
NIH 3T3, 30
Nucleosome free region, 275

One gene-one cancer hypothesis, 21
Oncogenes 13, 23, 79, 80, 97, 186, 209
    c-*myc*, 66, 69, 130, 162
    fps, 27, 80, 107
    H-*ras*, 80
    K-*ras*, 80
    myb, 27
    N-*ras*, 80
    proto-*mos*, 30
    proto-*myb*, 27
    proto-*myc*, 22
    v-*abl*, 80, 107
    v-*erb* A, 80
    v-*erb* B, 80
    v-*fes*, 80
    v-*fgr*, 80
    v-*fms*, 80
    v-*fos*, 80
    v-*mil*, 80
    v-*mos*, 80
    v-*myb*, 80
    v-*myc*, 66, 80, 130
    v-*raf*, 80
    v-*rel*, 80
    v-*ros*, 80
    v-*sis*, 80
    v-*ski*, 80
    v-*src*, 80
    v-*yes*, 80

Oncogenic function, 47

p36, 83, 106, 210
p53, 119
PAH tumorigenesis, 290
Papillomas, 39
Papova viruses, 135
Parvoviruses, 175
    autonomous, 177
    immunosuppressive effect, 176
PCNA, 210, 212, 223
P450 dependent enzyme systems, 290
PDGF, 83, 196
Phenobarbital, 292, 294
Phosphopeptide mapping, 102
Phosphoproteins, 97
Phosphoserine, 98
Phosphotyrosine, 81, 84, 98, 99
Phosphotyrosine antibodies, 97
plt gene, 129
Plasma cells, 68
Plasminogen activator, 121
pmt gene, 129
Polyoma, 268
Polyoma virus, 127, 136
pp60V-*src*, 81, 102, 103
Proliferating cells, 223
Promoters, 267
Promotion, 22, 291
Protein data base, 210
Protein-serine/threonine kinases,
    86
Protein-tyrosine kinase, 81
Proto-*onc* genes, 21, 147
PTLV, 191

QUEST software, 211
Qualitative model, 31
Quantitative model, 31
Quiescent state, 209

Rat hepatomas, 242
Rearrangement, 36
Reduction of tumors, 178
REF 52, 114, 210
Regulatory gene, 170
Retroviruses, 21, 97, 185
Rous sarcoma virus (RSV), 24, 97
Reverse transcriptase, 187

Sarcoma viruses, 30
    Fujinami, 24
    Harvey, 30
    Kirsten, 24
    Moloney, 30
    Rous, 24
Second large T-antigen exon, 114
    transformation capacity, 116
Serum lot sensitivity, 10
Sezary syndrome, 186
Single-gene determinants, 23
Single stranded DNA, 175
Somatic cell hybrids, 68
S-phase, 227
    early, 229
    late, 229
    subdivision, 227
Spontaneous transformation
    frequency, 4
SV40, 113, 136, 210, 268
    DNA, 113
    DNA fragments, 113
    large T-antigen, 113
    oncogenicity, 113
    small T-antigen, 113
SV40 DNA, 113
    microinjection, 113
Syncitia formation, 188
Syrian hamster embryo (SHE), 2
Syrian hamster embryo-simian
    adenovirus (SA-7), 5
T-antigen, 113, 128, 137
    large, 113, 128, 142
    middle, 128, 141
    positive nuclei, 114
    small, 113, 128
    synthesis and maintenance of
        transformed state, 116
T-antigen specific functions, 114
    cell DNA synthesis, 114
    helper function for adeno
        2 virus, 114
Target cell pools, 10
TATA box, 116
T cell growth factor, 186
    IL-2, 186
    TCGF, 186
TCGF receptor, 193
Terminal differentiation, 256
T-lymphocytes, 177

Transcription, 167
Transcriptional control, 244, 268
Transcription factors, 277
Transfection, 32
Transformation, 1, 65, 79, 113,
    127, 135, 157, 167, 175, 194
    209, 216, 239
Transformation frequency, 179
Transformed fibroblasts, 97
Transforming proteins, 97
Transgenic mice, 47
Translocations, 160
Tropomyosin, 212
Tubulin, 212
Tumor promoters, 291

Tumor promotion, 8
Tumors, 147
Tyrosine kinase activity, 81
Two-step transformation, 129
Two-dimensional gel electro-
    phoresis, 210

Vaccine, 199
Vimentin, 212
Vinculin, 83, 103
    antibodies, 105
Viral mutants, 142

Yolk sac
    alpha-fetoprotein, 239